New D...

an...

Genetic D...ses

New Biology
and
Genetic Diseases

BAKHTAVER S. MAHAJAN
Homi Bhabha Centre for Science Education
Tata Institute of Fundamental Research
Mumbai, India

MEDHA S. RAJADHYAKSHA
Department of Life Sciences
Sophia College for Women
Mumbai, India

OXFORD
UNIVERSITY PRESS

OXFORD
UNIVERSITY PRESS

YMCA Library Building, Jai Singh Road, New Delhi 110001

Oxford University Press is a department of the University of Oxford. It furthers the
University's objective of excellence in research, scholarship, and education
by publishing worldwide in

Oxford   New York
Athens Auckland Bangkok Bogota Buenos Aires Calcutta
Cape Town Chennai Dar es Salaam Delhi Florence Hong Kong Istanbul
Karachi Kuala Lumpur Madrid Melbourne Mexico City Mumbai
Nairobi Paris Sao Paolo Singapore Taipei Tokyo Toronto Warsaw

with associated companies in

Berlin   Ibadan

ISBN 0 19 564769 6

Illustrations by:
Anand Ghiasis
Maithili Rode
Medha S. Rajadhyaksha

Typeset by Sheel Arts, Delhi 110 092
Printed in India at Rashtriya Printers, Delhi 110 032
and published by Manzar Khan, Oxford University Press
YMCA Library Building, Jai Singh Road, New Delhi 110 001

CONTENTS

CONTENTS

FOREWORD

Genetics is a relatively new science. It properly began with the rediscovery of Gregor Mendel's work at the commencement of the twentieth century. In 1900 Landsteiner found that the blood of people could be classified in three groups, A, B and O. Landsteiner's discovery had a simple Mendelian interpretation. At about the same time, Archibald Garrod, a British physician, discovered an inherited metabolic disorder called Alkaptonuria in which the patients' urine turned dark. Garrod surmised that the darkening of urine was due to an inborn error of metabolism caused by a defective enzyme. This was the first suggestion that genes make enzymes.

Although human genetics had an early and brilliant beginning, its progress was slow because experimental methods of Mendelian genetics were not easily applicable to humans. Geneticists depended upon the analysis of gene frequencies in populations and transmission of traits in families. The situation has now changed dramatically. With the advent of molecular genetics, study of Mendelian genes has become the study of DNA in test tube. Human genetics is now more advanced than the genetics of most other animals. A complete physical map of human chromosomes has been made. It is expected that in another five years or so the entire DNA sequence of humans will be known. All genetic information in our nucleus will be available as an open book. This will be a scientific achievement of unprecedented import which will revolutionize our understanding of human genetics. It will also greatly improve our ability to diagnose and deal with inherited genetic diseases.

Bakhtaver Mahajan and Medha Rajadhyaksha have written a book on inherited diseases for the uninitiated. They present a very readable account of important genetic disorders, together with a background knowledge of genetics and molecular biology necessary for the subject. The book covers a variety of inherited disabilities ranging from blood diseases to mental illness.

Genetic disabilities afflict all sections of our population. Small groups, specially those that practise customary inbreeding, carry a heavy load of certain diseases. The incidence of thalassaemias, for instance, exceeds 30 per cent in certain tribal groups. We urgently need easily accessible genetic counselling and diagnostic help, at present available to only a few in cities.

Human genetics as a subject is greatly neglected in our education, specially medical education. I hope this book will be widely read. It should increase public awareness in an area which is often shrouded in ignorance and prejudice.

O. Siddiqi
National Centre for Biological Sciences
Bangalore

PREFACE

New Biology and Genetic Diseases is a popular science book born out of our experiences with students at the school and college levels. These experiences convinced us that most students, even those with formal training in biology, had poor understanding about the inheritance of different characters, including diseases, and their genetic basis.

Certain inherited diseases, such as the blood disorder thalassaemia and a wide range of muscular dystrophies, have high prevalence in the Indian population. The numbers of those affected run into lakhs. Every year, about 10,000 thalassaemic children are born in India! In addition to these two diseases, there are several others—some with little or no consequences, and others with devastating clinical symptoms leading to early death—which are genetic in origin. They inflict great pain and misery on the affected individuals and their families. The custom of marrying within family and caste groups, widely practised in India, seems to further aggravate this problem. What is not normally realized is that the insights offered by recent advances in genetics can prevent several of these diseases, and in some, even cures are being attempted.

The knowledge about inheritance of several traits in humans, animals and plants is very old. This knowledge has been used in domestication of animals, in agriculture, and in breeding of better varieties of animals and plants. But genetics as a science is a young discipline. Gregor Mendel's experiments with peas, published in 1865, are generally accepted as the first genetic experiments. However, it was Archibald Garrod's prescient observations about inherited diseases and his classical book, *Inborn Errors of Metabolism*, in 1902 which drew attention to human genetic diseases and their high incidence whenever the parents were first cousins. Since then, and especially in the last three decades, an unbelievable progress has been made in this area. What to Mendel were purely abstract heritable factors which determined specific traits in all living organisms are today known as genes whose chemical structure and function have been dissected to the finest detail.

It is now recognized that heritable diseases are due to defective structure or function of specific genes. The information content of all the genes of human beings, contained in the nucleotide sequences of DNA, will be fully known by 2005 or even earlier. What is more, today genetic changes responsible for specific diseases can be determined precisely, and the hereditary transmission of such defects can be prevented by genetic counselling to individuals with such defects. In some cases, cures may be possible by the supply of a normal (functional) copy of the affected gene. The knowledge of all these developments can greatly reduce individual human suffering, as well as the economic and social costs to society. Simultaneously, this deluge of genetic information also poses several ethical, legal, and social dilemmas which must be resolved by broad discussions at the popular level. That will be possible only if the common people have the knowledge of at least the basics of this field. We hope this book will be a small step in this direction.

The book covers a broad canvas and has a historical approach. The introduction, rather than having a continuous thread, attempts to give the reader a flavour of what is in store for humanity as the Human Genome Project nears completion. Already there are several sites on the World Wide Web giving information about a large number of genes—not only of humans, but of a variety of organisms, ranging from different bacteria and yeast to animals, such as worms, flies, fish, rats and monkeys. These data and genomic maps, a term broadly used for different genetic and physical maps, have started to give us a much more coherent view of life. They reveal a profound commonality among the genetic blueprints of different forms of life. At the philosophical level, this knowledge will reorient our sense of the self and weaken the ethnocentric world-view. Humans are just another living organism, evolutionarily one of the youngest. The myth of an exalted animal created specially to rule over and exploit the rest of nature lies truly buried forever. Ironically, this is at a time when special faculties unique to humans have enabled them to discover the truth about themselves and to acquire the power to mend, mix and mould all organisms according to their needs and wishes. At a mundane level, these advances in genetics are finding great uses in medical sciences, especially in the prevention of genetic diseases. Not surprisingly, this compilation, analysis, and application of our genetic data is considered as one of the greatest accomplishments of our times.

Bakhtaver S. Mahajan
Medha S. Rajadhyaksha

ACKNOWLEDGEMENTS

In bringing out this book, we have received help from several quarters. First, we would like to acknowledge with deep gratitude the work put in by Dr M. K. Gupta, formerly with the Bhabha Atomic Research Centre (BARC) as the Associate Director of the Electronics and Instrumentation Group, who painstakingly read the book and came up with numerous suggestions to make it more readable for non-biologists and lay persons. Dr D. S. Joshi, working in the field of human genetics at the Molecular Biology and Agriculture Division of BARC, checked the contents for possible technical errors and gave us several suggestions. We highly appreciate their help.

Gracious support in the form of supplying us original photographs and case studies was given by several scientific organizations and individuals associated with human genetics. We would specially like to thank Dr Sudha Gangal, Director, Research Society, B. J. Wadia Hospital for Children and Institute for Child Health, Mumbai, and Mrs Harsha Yagnik, genetic counsellor with the same organization, for their guidance and for giving us several photographs and pedigrees of those affected with thalassaemia. Original photographs of a normal karyotype (p. 51) and of a patient afflicted with fragile X syndrome (p. 55) were obtained from Smt Motibai Thackersey Institute of Research in the Field of Mental Retardation, Sewri Hill, Mumbai. Mrs Saroj Iyer, Honorary Principal of Sausheelya Special School at Anushaktinagar and Dr Varsha Vadera, Chairperson of the Project Thalassaemia of the Rotary Club of Mid-Town, Mumbai, also gave several inputs and photographs.

The receipt of a greeting card exquisitely painted by Anand Bantwal of Goa prodded us into taking a different perspective of those afflicted with inherited diseases. These cards (p. 60) are meant to raise funds for the Indian Muscular Dystrophy Association. While Dr Appaji Rao of the Institute of Science, Bangalore, readily sent his publications on consanguineous marriages which helped considerably in clearing certain doubts, Dr Noshir Wadia of the Jaslok Hospital, Mumbai, spent time with us answering some of our questions on inherited childhood problems. We are indebted to all.

The book has a distinct Indian character thanks to several imaginative drawings made by Anand Ghiasas of HBCSE. Maithili Rode of the J. J. School of Arts and Architecture made the line drawings of the illustrations, further improved by one of us. Colleagues at the HBCSE and Sophia College have willingly helped us at various stages. Vaibhav Palekar and Amrita Patil made several computer graphics and Ravi Patwardhan expertly formatted and printed the pages. Gajanan Mestry, Sumana Amin and N. Thigale willingly provided office assistance in xeroxing, typing, reducing the illustrations, and performed several other odd jobs. Dr Arvind Kumar, Director, HBCSE, gave us valuable suggestions and constant encouragement.

Lastly, both of us owe a lot to our family members who have bravely borne the brunt when the two of us were slogging over our book. We have enjoyed this experience of working together and look forward to other fruitful collaborations.

1

INTRODUCTION
What is this book about?

This is the story of the Vasvani family (the name is changed for reasons of confidentiality). This family, originally from Sind, is now spread all over India. What marks them out is that numerous members of this family, like thousands in the country, are capable of passing on to their children thalassaemia, a potentially debilitating and lethal blood disorder. Thalassaemia is an inherited disease.

Artist's impression of the Vasvani family.

This is also the story of **human genetics** which will soon usher our society into a new age. This will be an era where several of us are going to be affected as we learn more about our inheritance, especially about the hereditary molecule, deoxyribo nucleic acid, popularly called DNA. The chemical structure and organization of this molecule allows it to carry information for all our traits, in discrete portions called genes. DNA is present in all our cells[1] and is passed on from parents to their children.

Advances in human genetics are moving at a breathtaking pace in the world. Several Indian laboratories too, are active in this discipline.

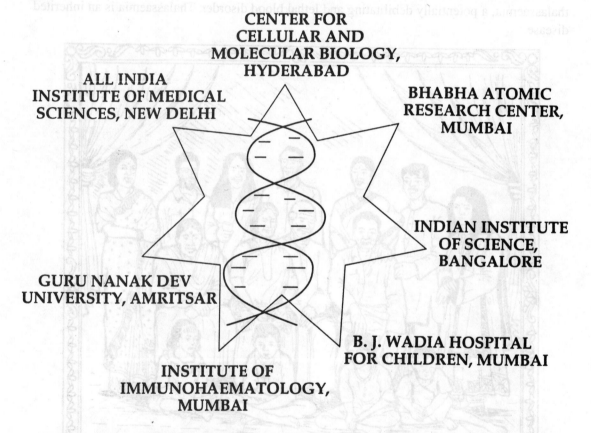

CENTER FOR CELLULAR AND MOLECULAR BIOLOGY, HYDERABAD

ALL INDIA INSTITUTE OF MEDICAL SCIENCES, NEW DELHI

BHABHA ATOMIC RESEARCH CENTER, MUMBAI

INDIAN INSTITUTE OF SCIENCE, BANGALORE

GURU NANAK DEV UNIVERSITY, AMRITSAR

B. J. WADIA HOSPITAL FOR CHILDREN, MUMBAI

INSTITUTE OF IMMUNOHAEMATOLOGY, MUMBAI

The venture to decipher the information in our genome, in the total DNA of our cells, under the aegis of the Human Genome Project (HGP), is nearing completion. Soon, it will be possible to screen all the information in our genome at the click of a computer mouse.

Identification of the exact regions of the DNA which carry information for making different proteins is no more in the realm of science fiction. And it is the protein molecules that are expressed as different traits such as the colour of our eyes, strong and weak bones, big or small

---

[1] Mature red blood corpuscles (RBCs), which do not have a nucleus, are an exception.

nose, and several disabilities and diseases, including thalassaemia. With most of the primary structure of our DNA deciphered, it would be possible to diagnose inherited diseases at an early stage, even before the birth of a baby.

It is now possible to examine the DNA of a fertilized egg in its early stages of multiplication and implant only the 'healthy' pre-embryo into the womb, especially where there is a family history of an inherited disease. Trials are also under way to find possible treatments and cures by gene therapy for several genetic disorders. Better molecular understanding of complex diseases, such as arthritis and heart problems, could lead to strategies for their prevention and possible cures.

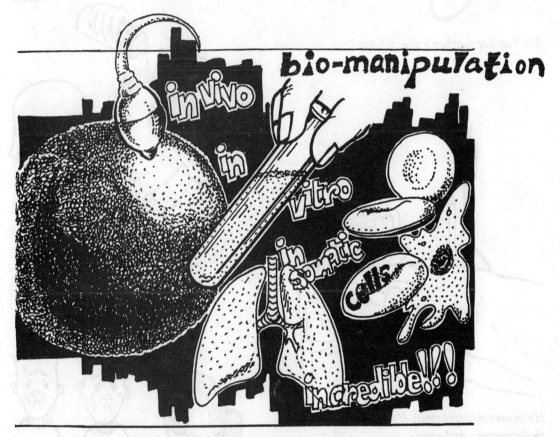

Several types of human behaviour, including alcoholism and depression, have genetic roots, though the role of environment cannot be ignored. A variety of mental traits, such as anxiety, memory and even our sleep patterns, are today being examined for possible genetic components.

The use of a molecular approach in the treatment of mental diseases is high on the agenda of psychiatrists and those dealing with behavioural problems. Molecular medicine is the name of the new discipline. In future, clinicians will not be able to ignore the science of molecular genetics.

She has her mother's sharp nose.

Of no use at all!
He has his father's bald pate.

He is severely anaemic. He has
thalassaemia and needs
frequent transfusions.

Both the siblings are eternally
depressed. So was their mother.
May be she created a depressive
environment.

Advances in human genetics will enable physicians and genetic counsellors to inform us of our predisposition to specific inherited diseases, certain behavioural defects and our susceptibility to certain communicable diseases! Indeed, genes do determine a lot about diseases of an individual.

These developments are likely to raise several ethical questions and major dilemmas for us.

Do the roots of anti-social behaviour lie solely
in our genes? NO! Criminally aggressive
behaviour could be traced to gene-
environment interactions.

Will abortion of the foetuses with 'unwanted'
genes improve society? This is a
complex question. However, by doing so, there
will be reduction in the burden on
individual families and
costs to the society.

Several controversial legal issues of
gene-based patents are bound to crop
up as the human genetic heritage is examined
in its minutest molecular detail. Drugs developed
based on your own DNA will not be available to you at
affordable costs.

Some activists view the developments
in human genetics as an affront to
individual dignity and
identity!

It is not all in the genes alone !

What are the new questions and possible dilemmas which will crop up with advances in human genetics? Consider the following hypothetical situations:

(1) 'My mother and sister both died of breast cancer. Does this imply that I will also develop the disease? What about my two daughters? Are they also at risk of developing breast cancer?'

What if the two daughters are not tested for, say, the breast cancer genes? Does a negative test really guarantee one a life-time protection against cancer? Or, does the presence of a disease gene always lead to the development of the corresponding disease?

(2) 'My paternal grandmother was confined to a room as she was considered to be mentally retarded. My father has strong streaks of depression and violent behaviour. What does this mean for me? Am I at risk too?'

But the environment of the child is entirely different from that of her father and grandmother.

(3) 'All the members of my family are obese, including those on a strict diet regime.'

Is obesity a genetic trait?

(4) 'I am tired of looking after sick members of my family. My first child is thalassaemic.'

Is it not possible to have a baby with 'healthy' genes?

These and several other questions will soon be answered by undertaking genetic tests, involving a few drops of blood or some cells from our body.

Tests for diagnosing several inherited diseases are already available and new ones for more diseases are being continuously designed. This process is greatly accelerated as new genes are rapidly being discovered. But these very developments may also pose problems for us.

Some other questions which could be raised are:
Who should have access to information about our genetic make-up?

SHOULD THIS DATA BE MADE AVAILABLE TO ...

(1)   *...all family members? What about young adults?*

The knowledge of the impending disease, its inevitability and lack of cure might encourage fatalistic, even suicidal tendencies. These persons need psychological help to handle such knowledge.

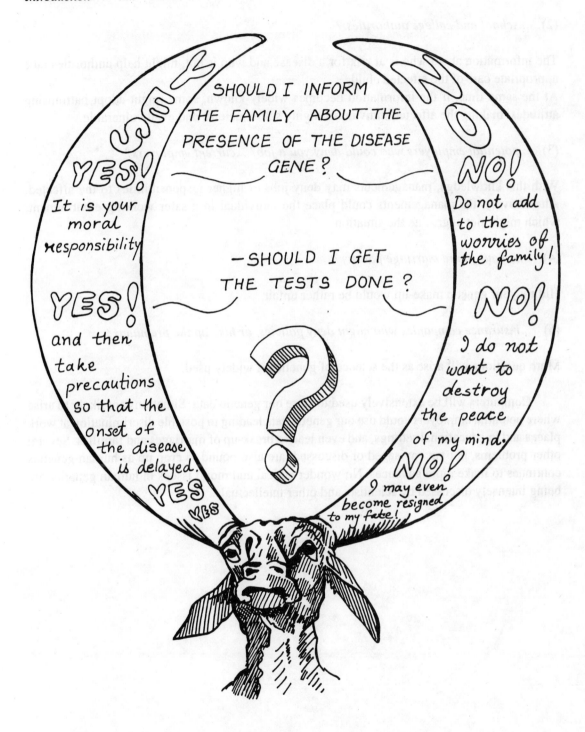

(2)   *...school and college authorities ?*

The information about who is at risk for a disease and who is not, might help authorities take appropriate care of the affected child.
At the same time, if the information becomes widely known, some might adopt patronising attitudes, making the affected individuals highly self-concious, and even vulnerable.

(3)   *...potential employers who could deny you a job? ...current employers?*

With this knowledge, managements may deny jobs or higher responsibilities to the affected. Alternatively, the managements could place the individual in a safer working environment, which might not aggravate the situation.

(4)   *...to your future marriage partner?*

Hiding your genetic make-up would be rather unfair.

(5)   *...insurance companies who might deny policies, or hike up the premiums?*

More questions will arise as the science of genetics is widely used.

Computers will be extensively used to store our genetic data. Situations are likely to arise where potential employers could use our genetic data leading to possible discrimination at work places and in educational settings, and even lead to break-up of marriages and families. Several other problems, not yet envisaged or discussed, are also bound to crop up as human genetics continues to make rapid advances. No wonder ethical and moral issues in human genetics are being intensely discussed by scientists and other intellectuals.

2 BASIC IDEAS IN GENETICS
At the cellular and molecular levels

This book also narrates how human genetics blossomed. **Genetics** is the science of heredity and variations in all living organisms. Basic understanding in this area will help us greatly to take right decisions and formulate suitable answers to complex questions concerning human genetics.

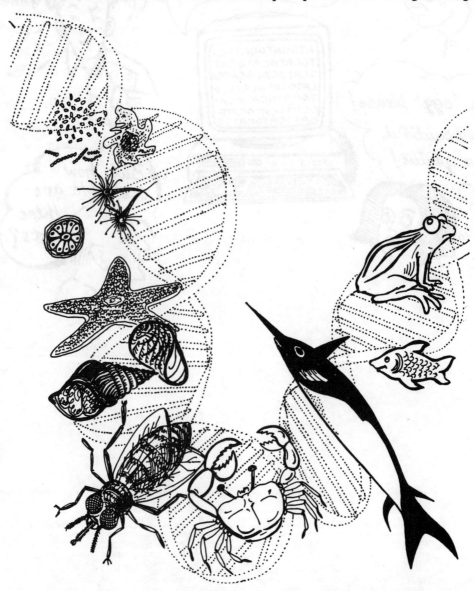

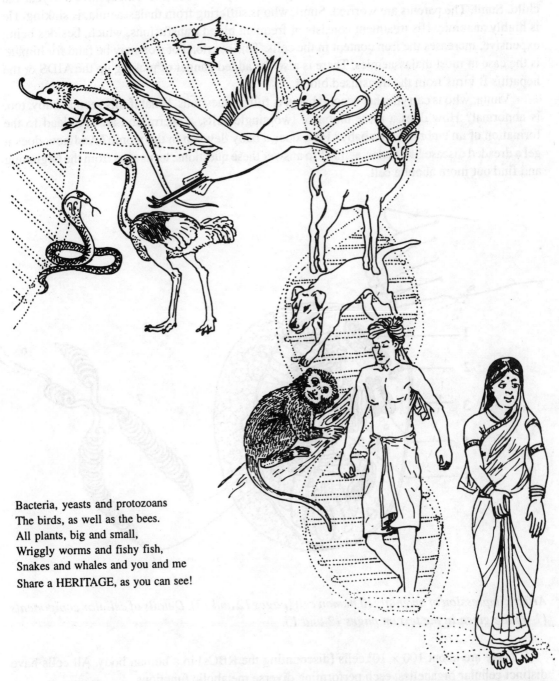

Bacteria, yeasts and protozoans
The birds, as well as the bees.
All plants, big and small,
Wriggly worms and fishy fish,
Snakes and whales and you and me
Share a HERITAGE, as you can see!

All living organisms share a common origin, a common biochemistry, including the genetic apparatus and the genetic code. What a fantastic unity in this diversity!

Let's come back to the Vasvanis. The young couple, Vinita and Vinod, have a 3-year old child, Sunil. The parents are worried. Sunil, who is suffering from thalassaemia, is sinking. He is highly anaemic. His treatment consists of frequent blood transfusions, which, besides being expensive, increases the iron content in the cells. This overload of iron may be fatal for him, as is the case in most thalassaemics. There is also the added danger of picking up the AIDS or the hepatitis B virus from the transfused blood.

Vinita, who is carrying her second baby, is full of questions. What if this second baby, too, is abnormal? How does a mere union of two single cells, a sperm and an egg, lead to the formation of an embryo and a baby? How does a baby develop in the womb? And how does it get a dreaded disease like thalassaemia? To answer these questions, let us start from the beginning and find out more about a cell.

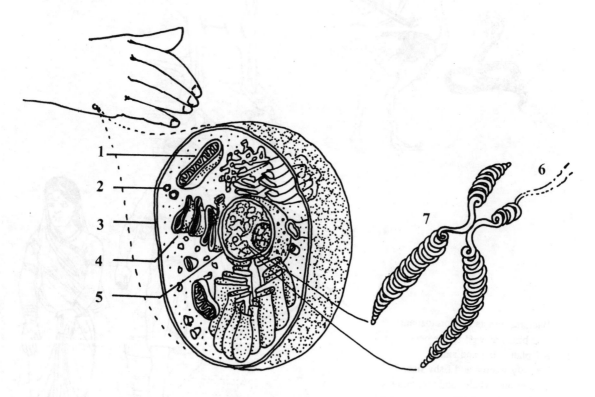

Artist's impression of an enlarged human cell (pages 12 and 13). Details of cellular components (1–7) are given in the text on pages 12 and 13.

There are about 100×10^9 cells (discounting the RBCs) in a human body. All cells have distinct cellular organelles, each performing diverse metabolic functions.

1. Mitochondria are the power house of a cell.

2. Ribosomes help in protein synthesis.

3. The cell membrane forms a complex boundary of a cell.

4. Golgi complexes pack and modify all that goes out of a cell.

5. Nucleus is the most prominent part of a cell and contains DNA.

6. DNA molecules carry the genetic information as sequences of chemical bases—genes—capable of synthesizing proteins which run the cellular machinery. Any defects in this information, in the arrangement of nitrogenous bases in the DNA molecule, can lead to defective proteins and to inherited diseases like thalassaemia.

7. As a cell readies to divide into two equal halves, the DNA, along with some special proteins, organizes into compact, rod-shaped bodies called chromosomes. The newly formed chromosomes are divided equally between the daughter cells which thus inherit all the properties of a parent cell. These chromosomes are passed from parents to offsprings. We have 23 PAIRS (46) of chromosomes in our cells, except in male and female reproductive cells, eggs and sperms. Of these, one pair determines the sex of an individual—two X's in a female (XX) and one X and one Y (XY) in a male. These are the sex chromosomes. The other chromosomes are known as the autosomes.

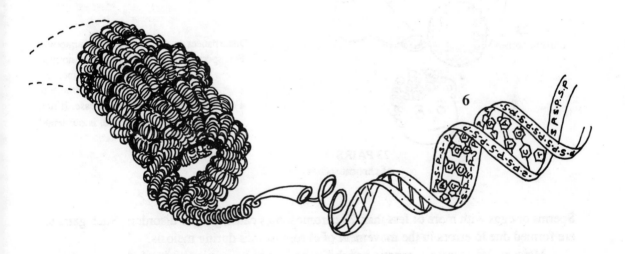

The reproductive cells or gametes, i.e., sperms and eggs have 23 chromosomes, only one member of each chromosome pair. This reduction in the number of chromosomes takes place by a special type of cell division, meiosis. This process occurs solely in the reproductive organs.

The sperms are produced in the testes throughout the reproductive life of a man. About 300 million sperms are released during each ejaculation. Meiotic division takes place in the male testes during sperm production. In contrast, meiosis has an early start in female babies with the formation of immature eggs or ova in the ovaries, even before their birth. But these eggs are in a state of suspended animation, though the reduction in the number and shuffling of chromosomes has already taken place at this stage. One or few eggs are released every month starting from puberty and lasting for the next 20 to 30 years. Meiosis gets completed only after fertilization.

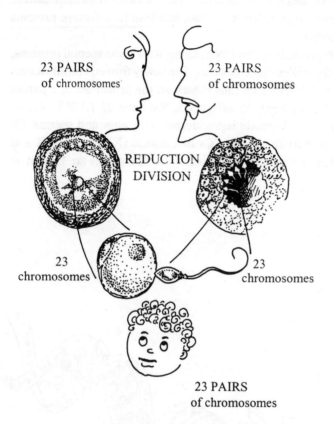

23 PAIRS
of chromosomes

23 PAIRS
of chromosomes

REDUCTION
DIVISION

23
chromosomes

23
chromosomes

23 PAIRS
of chromosomes

Disturbances in meiosis, an affair abnormal
For, sperms and eggs result, not so normal.
Fertilization still goes on,
But foetus not further on.
For aborted foetus is often the case. If not,
Congenital disease, sure, is our bane!

Sperms or eggs with more or less than 23 chromosomes cause genetic disorders. Such gametes are formed due to errors in the movement of chromosomes during meiosis.

Meiosis also increases genetic variability and gives us a unique mix of characteristics of our parents. Let's see how this happens.

Early meiosis is characterized by several important events, many of which still continue to bewilder scientists for their precision and sheer beauty. First, there is the condensation of the thready DNA mass into chromosomes, followed by pairing of identical (homologous) chromosomes, known as synapsis. After the chromosome pairs are neatly aligned, there is a breakage and exchange of parts between two of the four homologous DNA threads. This is known as crossing over. An important aspect of crossing over is its fairly random occurrence along the length of the chromosomes. On several occasions, crossing over may not occur, yielding germ cells having parental genes. Alternatively, crossing over between homologous chromosomes causes shuffling of parental genes in a given chromosome pair, giving rise to germ cells with some genes from one or the other parent.

'How does this happen?' How does the union of a sperm and an egg produce so much variety?' 'How does it all start?' Others, long before Vinita, have been asking such questions. Now we know that the blueprint for growth and development of a baby is contributed by the DNA present in the germ cells.

A healthy and fertile man releases millions of sperms in an ejaculation, of which about 100 attach to an egg, but only one manages to enter the egg, forming a zygote. Improved microscopy has revealed the detailed structure of sperms. Each has a nucleus hugging a cap-like structure, a short neck and whip-like tail.

Man produces a large number of malformed or immotile sperms. Some of the physical abnormalities observed in sperms are: immature nucleus with uncoiled and uncompacted DNA, a malformed or absent cap (acrosome), an abnormally long neck, extra tails or completely double sperms. Not many abnormal sperms are able to swim their way into the female reproductive tract, but some can travel into the fallopian tube and even fertilize the egg. Expectedly, these sperms do not produce a normal zygote.

Of about 2 million immature eggs present in the ovaries of newborn female babies, about 30,000 to 40,000 remain viable till puberty. And of these, only 400 gain maturity and are released every month starting from puberty. This prolonged maturation of the eggs in the ovary —spanning several years—makes them highly susceptible to environmental influences. Chances of chromosomal abnormalities occurring in them cannot be ruled out. No wonder, the incidence of abnormal babies is higher in mothers who are well above 35 years old than their younger counterparts. Just as the mother's age increases the chances of abnormality in the babies, older fathers too accumulate changes in the genetic material of their sperms which the baby is likely to inherit.

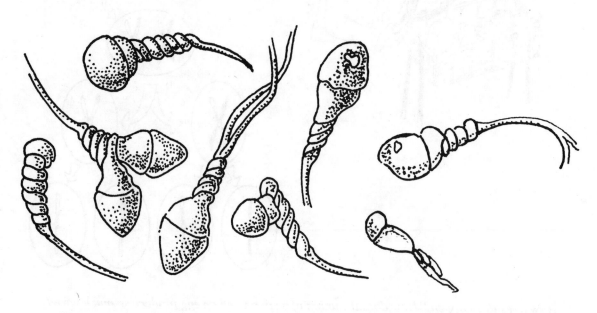

Defective sperms: A man is as much to be blamed as a woman for genetic abnormalities in a baby.

A woman becomes pregnant with the union of a sperm and an ovum in the fallopian tube. The sperm nucleus gets gobbled up by the ovum in a few seconds, and then the ovum becomes impenetrable to other sperms. The two germ cells, each with 23 chromosomes, unite to form a zygote, containing 46 chromosomes (23 pairs). The entire process of fertilization takes less than 20 minutes.

Development now proceeds rapidly, with the zygote dividing and redividing by mitosis into a two-celled, four-celled and several celled structure, as it travels from the fallopian tube to the uterus.

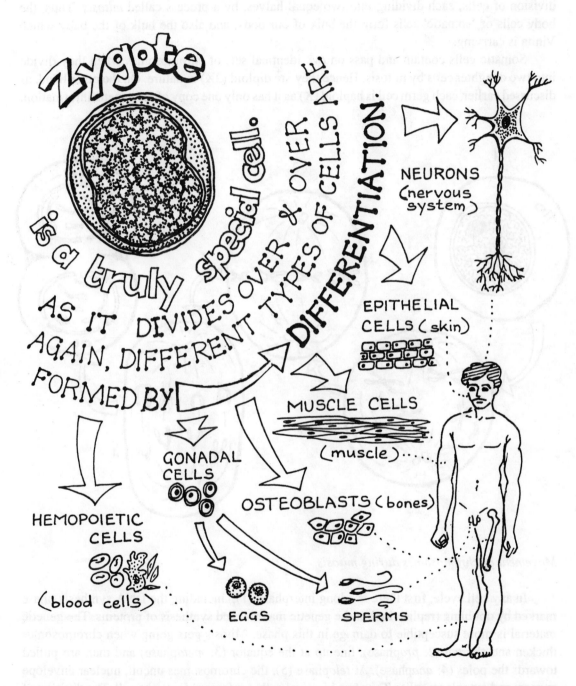

Growth and differentiation are highly complex processes sensitive to genetic and environmental factors.

The growth of a baby from the stage of the fertilized egg to an adult takes place by rapid division of cells, each dividing into two equal halves, by a process called *mitosis*. Thus, the body cells or 'somatic' cells form the bulk of our body, and also the bulk of the baby which Vinita is carrying.

Somatic cells contain and pass on *two* identical sets of chromosomes when they divide into two daughter cells by mitosis. Hence they are diploid (2X) in nature. On the other hand, as discussed earlier, each germ cell is haploid (X) as it has only one copy of the genetic information.

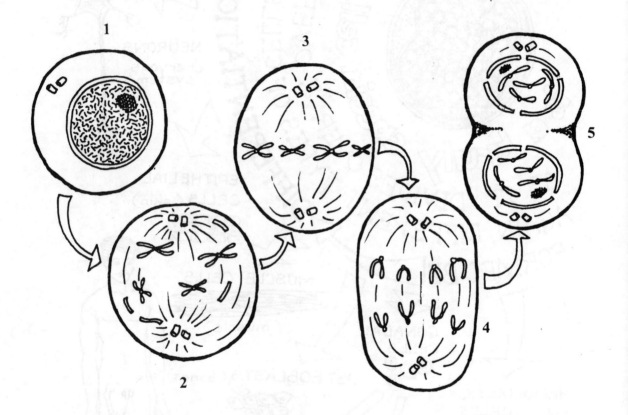

Movement of chromosomes during mitosis.

In any cell cycle, first there is a long interphase (1), including the DNA synthesis phase marked by doubling (replication) of the genetic material and synthesis of proteins. The genetic material is more susceptible to damage in this phase. Mitosis gets going when chromosomes thicken and shorten (2: *pro*phase), line up at the equator (3: *meta*phase) and then are pulled towards the poles (4: *ana*phase). At *telo*phase (5), the chromosomes uncoil, nuclear envelope appears and cytoplasm splits. Thus two identical cells are formed from one cell. The division of a cell into two daughter cells is one of the most amazing events taking place in living organisms.

Cell division by mitosis takes place continuously in humans, though it slows down considerably in adults. Mitosis is somewhat rapid during repair of injuries, or during healing of wounds, resulting in replacement of injured cells. However, there is no restoring of certain cells, such as muscle and nerve cells, once they are formed. Injury to either could lead to serious consequences. Other cells, such as the skin cells on the surface or the blood cells, are constantly dividing. This makes them more prone to genetic insults.

Foetal cells, too, are dividing rapidly and hence are susceptible to genetic anomalies.

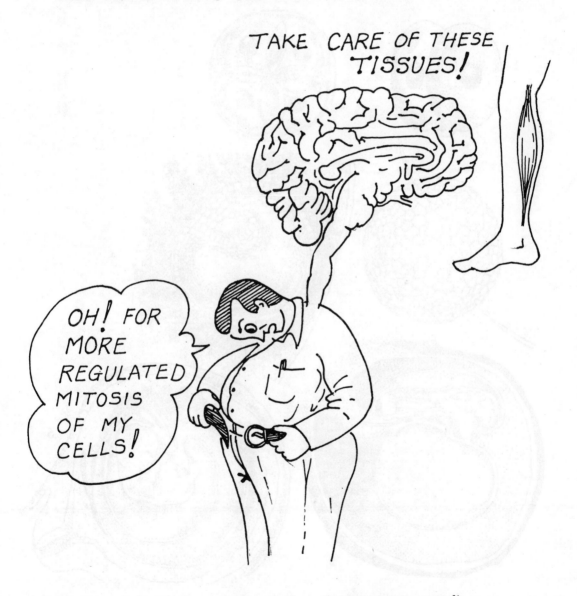

Mitosis is not responsible for your paunch. The cell volume has increased!

The mitotically dividing zygote tightly latches on to the uterine wall and slowly begins to grow and differentiate into different tissue-cells and organs of the foetus. Growth and development of a zygote into an embryo involves selective functioning of the genes in different body cells. Some genes are switched on and others switched off, affecting their ability to form a protein product in different cells of the tissues. Hence the cells which go to form the liver have a specific set of functioning genes and those forming the skeletal muscles have others. The results could be disastrous if there are any disturbances in gene action in these differentiating cells.

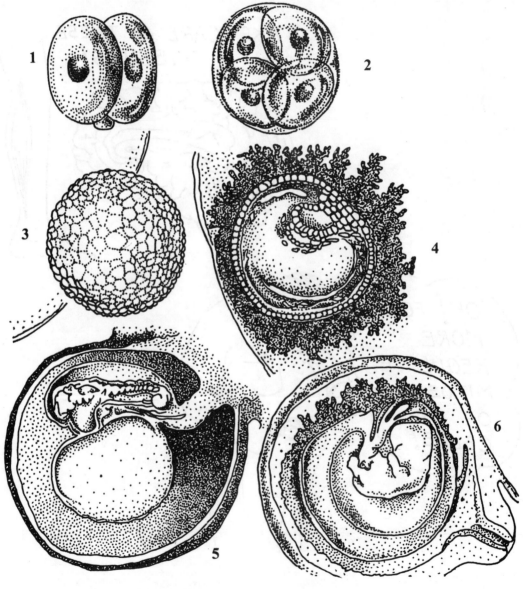

Different stages of early development of a human baby (figs. 1-6).

The eighth week of pregnancy marks the end of embryonic stage. By this time all the essential organs—heart, brain, liver, kidneys, lungs, etc.—are formed. The foetal period starts from the ninth week till birth and is characterized by growth in size and further elaboration of the organs which are already formed.

Questions about development of an embryo continue to fascinate and baffle scientists. There are still no complete explanations about how differences arise in cells whose nuclei, to start with, were genetically identical!

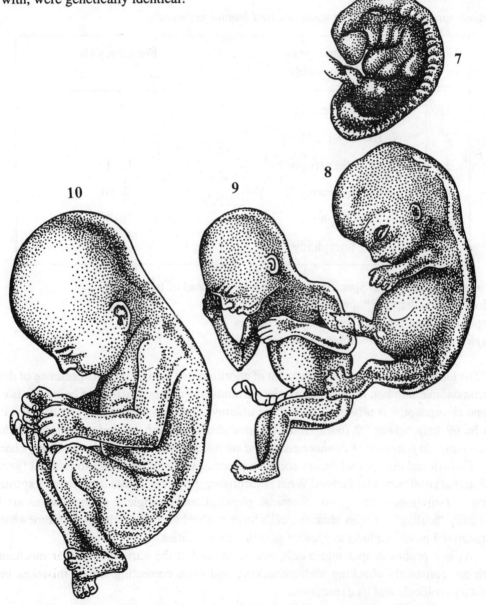

Some important stages of later development of a baby in the womb (figs. 7 to 10).

In most cases, fertilization and zygote development proceed smoothly to form a healthy baby. However, spontaneous abortions are known to occur and it is estimated that about 15 per cent of all confirmed pregnancies terminate in abortions. Indeed, examinations of the aborted foetuses have revealed a variety of abnormalities in the chromosome number and structure. Nature has evolved its own way of getting rid of abnormal embryos by spontaneous abortions, which might have given birth to babies with genetic abnormalities.

Chromosomal abnormalities in some aborted human embryos

	Chromosomal Abnormality	Frequency (%)
1.	Trisomy 16	7–8
2.	Trisomy 13, 18 or 21	4–5
3.	All other trisomies	13–16
4.	Extra or missing X or Y	9–10
5.	Triploidy	6–7
6.	Tetraploidy	2–3

Trisomy= three copies of specific chromosomes—instead of two copies
Triploidy= three haploid* sets of chromosomes
Tetraploidy= four haploid* sets of chromosomes
* *Single copy of 23 chromosomes as in germ cells.*

The table reveals that 40 to 50 per cent of abortions can be traced to the presence of defects in chromosomal number. In these cases, the chromosome number instead of being 46, is either 45 (one chromosome is missing), or 47 (one trisomy) or 48 (two trisomies). At times, it could even be 69 (triploid) or 92 (tetraploid). *Several diseases, such as thalassaemia, are due to abnormalities in genes which cannot be detected microscopically as gross chromosomal changes.*

Growth and differentiation are complex processes and are sensitive to internal (genetic) and external environmental factors! What is surprising, however, is that despite our exposure to a variety of environmental agents, chemical, physical, and biological, so many of us are born genetically 'healthy'. Besides abortion, cells have evolved several mechanisms (somewhat like computerized proof-reading) to prevent genetic abnormalities.

As one probes deeper into a cell, one is amazed at the varied molecular mechanisms which are constantly checking and rechecking and even correcting minor mistakes in our hereditary molecule and its expression.

Special types of enzymes, DNA polymerases, help in making new DNA molecules (replication). These enzymes also have another activity which can sense errors and remove them from the newly synthesized DNA strand.

Another set of enzymes has the capacity to recognize mismatches which arise due to incorporation of a wrong base or bases during replication. Any wrong insertion of a base in the newly synthesized DNA molecule is recognized, followed by its removal along with a neighbourhood stretch of the newly synthesized DNA. This stretch is replaced by a freshly synthesized piece of DNA.

Several other molecular mechanisms are also present in a cell which help in repairing DNA mismatches and other defects in the hereditary molecule.

> Molecular proof-readers,
> MutS and MutL,
> Have an extremely exacting
> Job to do!
> They patrol the DNA tip to tip,
> To check for any replication slip.
> They chew up and redo the molecular patch,
> That carries even a single base
> Mismatched!

There is considerable confusion about congenital diseases, which are defects present in infants at birth. Some of these are due to faulty genes inherited from one's parents that cause developmental defects in the growing embryo. Other congenital problems arise due to environmental factors. These diseases which are clearly visible at birth may or may not be transmitted. In contrast, inherited diseases are transmitted from parents to children and may appear at any stage of life.

Diseases, such as cleft lip and cleft palate, and spina bifida, are often erroneously categorized as inherited diseases. These are congenital diseases and occur due to developmental abnormalities taking place in the embryo. In cleft lip and palate there is an abnormal formation of the oral cavity; in spina bifida there is incomplete fusion of the spinal vertebrae, with resultant bulging of the nervous tissue out of its protective skeletal frame. Another example of a congenital abnormality which drew worldwide attention in the 1960s was the birth of a large number of babies with missing limbs during the 1950s and 1960s. Investigations soon brought out that the mothers of these babies were given a drug, thalidomide, to relieve anxiety and worry during pregnancy. The drug has since been withdrawn from the global market, but considerable harm was done to several thousands of people.

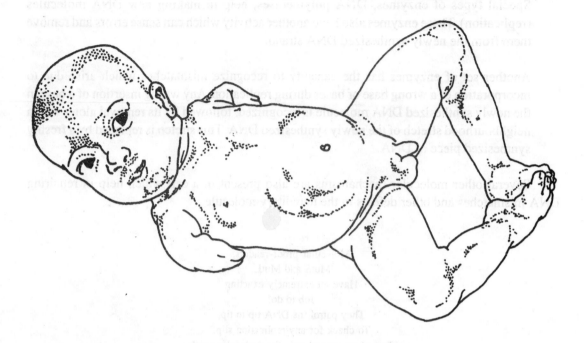

A baby with thalidomide syndrome.

Those affected by the thalidomide drug have typically short and deformed upper limbs and hence useless arms and hands. Often, the legs, ears, digestive tract, heart, and some blood vessels are also deformed. However, most of those afflicted have normal intelligence.

The developing foetus held snugly in the mother's womb, is cushioned by the wall of the uterus in the early days of pregnancy and by amniotic fluid in the later stages. The placenta is its lifeline which plays a dual role: it supplies the nutrients to the growing foetus and also protects the foetus from several harmful chemicals.

Still there are several substances from the environment which manage to cross the placental barrier and reach the foetus, and disturb its normal growth pattern. The substances which cause malformations in an embryo are called teratogens. The damage to the foetus is most severe if the exposure to the teratogen is during certain critical phases of its growth, especially when the cells are dividing rapidly during organ-formation. This period extends from the first week after fertilization to 3 months of pregnancy. In general, environmental disturbances in the first two weeks after fertilization can have drastic effects, even causing rapid expulsion of the foetus, resulting in abortion. From the 15th day to the 60th day, most organs are growing simultaneously and again, environmental exposure could interfere with their normal development.

The birth of babies with congenital defects has triggered research on what causes these abnormalities. This understanding has helped us to come up with some do's and don'ts for pregnant women.

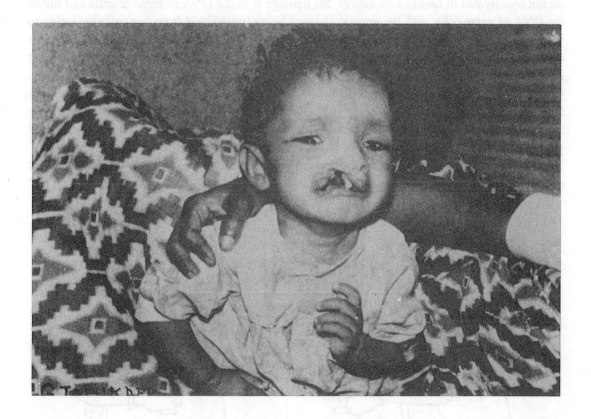

This baby has several congenital defects, including cleft lip & palate. Her parents, aged 35 and 46, are normal.

In addition to chemicals in the environment, several environmental factors can act as teratogens. These include alcohol, drugs and alkaloids. Hence, pregnant women are advised to avoid these. Strong medicinal drugs, including antibiotics, anti-convulsants and corticosteroids, should be taken under medical supervision. Oral contraceptives during early pregnancy have been implicated in foetal malformations. Tranquillizers and ingestion of chemical pollutants, such as mercury and lead, are teratogenic. Environmental ionizing radiation also affects foetal growth.

Exposure of pregnant women to several types of infectious agents could also turn out to be fatal. In this category, the culprits are the cytomegalo virus, herpes simplex virus and rubella virus (causing German measles). The effects on the foetus are more severe if the infection is in the early stages of pregnancy. The protozoa, *Toxoplasma gondii*, commonly known to infect food items and cause diarrhoea, can cause malformations of the foetal brain tissue; *Tryponema palladium*, which causes syphilis in adults, can lead to malformations in the baby.

What about cancer? Primarily the disease is genetic in nature, though most forms of cancers do not usually run in families. In cancer, the damage is in the DNA of somatic cells and not in the DNA of germ cells, with the genetic changes being transmitted from one cell generation to the next during the growth of an individual. There is a general agreement among cancer epidemiologists that this damage in the somatic cells is triggered by exposure to different environmental agents, and perhaps by an individual's life-style. The cancer-causing agents, carcinogens, act on specific regions of DNA, which we recognize as proto-oncogenes or tumour suppressor genes.

It is difficult to pin down the exact environmental causes of cancer, as it can take several years for clinical symptoms of cancer to appear after the exposure to a specific carcinogen. However, in cancers of the lung and the oral cavity, there is ample evidence of the involvement of nicotine and tobacco in triggering the disease.

Relaxing!! At what risks!!

How does exposure to different environmental agents, especially in the early months of pregnancy, cause developmental abnormalities in the growing baby? Ample scientific evidence indicates that complex interactions take place between these agents and the genetic material of somatic cells in the developing embryo, leading to physical and mental abnormalities.

It is important to recall that foetal cells are constantly dividing with the nuclear chromosomes undergoing replication. Hence the chances of these cells getting affected by environmental cues or signals are high.

Inherited diseases like thalassaemia, however, are not congenital. Thalassaemia, and other diseases of its ilk, have one or more defects in the hereditary molecule. With this understanding, one can well realize the anxiety of the Vasvanis.

The young Vasvanis are anxious about the fate of their second child. Their first child, Sunil, is a thalassaemic. They love Sunil, but still they would not like to be saddled with another diseased baby. Not knowing what to do, they seek the advice of a physician, who is known to counsel about inherited diseases.

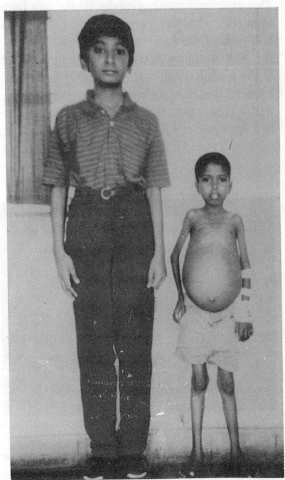

Both these children are 10 years old. The one on the left is normal; the child on the right has thalassaemia major.

This particular type of thalassaemia—β thalassaemia—or Cooley's anaemia, along with sickle-cell anaemia and several other blood disorders, causes severe anaemia among individuals. Till the 1980s, the affected babies died immediately after birth. But regular blood transfusions, often with some specific drugs, manage to give them a new lease of life. Some survive till their twenties or longer.

The severity of anaemia varies among those affected. At the extreme level, there is the lethal form of anaemia. Here an individual carries two copies of the genes for this disease—one copy each from the parents. At the other end, there is the mild anaemia which affects millions with no noticeable symptoms. Such people are the carriers and have inherited the trait from either one or the other parent. Generally, carriers lead relatively healthy lives. Both Vinod and Vinita are carriers.

Not knowing much about genetics or genes, the visits of the couple with the genetic counsellor were a revelation. The counselling opened up a 'brave new world' for them. The genetic tests revealed that Vinita's second baby carried two copies of the gene for thalassaemia. This gene-hunt was carried out by the new human genetics techniques, which are discussed a little later. Thus the Vasvanis started on a revealing journey, tracing the 'thalassaemia rogue trait' among the members of their families.

Genetic blood disorders, including thalassaemia, are caused due to certain structural defects in the protein (haemoglobin) present in the red blood corpuscles (RBCs). About 280 million haemoglobin (Hb) molecules are squeezed into each RBC, nearly filling it up. One can think of RBCs as bags of haemoglobin! Any defect in the structure of haemoglobin changes the structure and functions of RBCs, millions of which are coursing through our blood vessels.

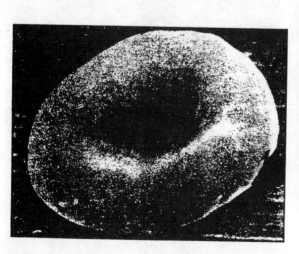

A normal RBC is round in shape (left). The one on the right is sickle shaped, as found in people afflicted with sickle-cell anaemia. See page 29.

RBCs with normal haemoglobin molecule (or with minor changes in the haemoglobin structure) have a spherical shape. Those with defective haemoglobin assume a sickle-like shape, as in sickle-cell anaemia, or are depressed as in different types of thalassaemias.

The determination of the structure of haemoglobin molecule represents a study of colossal order and complexity. Using X-ray crystallography and millions of computations, the 22-years' effort of Max Perutz and colleagues led to the determination of the positions of 10,000 atoms in this huge protein. It was, therefore, not a surprise when this effort was rewarded in 1959 with a Nobel Prize.

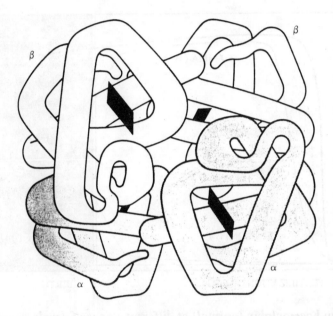

The 3-D structure of the haemoglobin molecule consists of the centrally located haem molecules encircled by the globin protein chains. The globin component consists of four convoluted chains—two alpha and two beta sub-unit chains. Each of the sub-units is built around the haem, on which sits the iron atom. As the RBCs travel to the lungs, these four iron atoms act as magnets and pull the oxygen molecules towards them, displaying incredible co-operation between the four sub-units. Once loaded with oxygen picked up from the lungs, the RBCs travel to different body tissues to unload this oxygen. This is done by an equally precise opening up of the globin chains. On the return journey, the haemoglobin molecule similarly releases the carbon dioxide picked up from the varied body tissues to the lungs, to be exhaled.

Alterations in the structure or quantity of any of the globin chains, changes the structure and functions of the RBCs, specifically its oxygen carrying capacity. Several structural defects are present in the globin molecule of thalassaemics.

Thalassaemia (from the Greek word, *Thalass-emia*= sea blood) is caused when the α or β polypeptide chains of the haemoglobin molecule are synthesized in reduced amounts.

Each person can make five different types of polypeptide chains of the globin molecule. These have been named after the first five letters of the Greek alphabet, viz, alpha (α), beta (β), gamma (γ) (of two sub-types), delta (δ) and epsilon (ε). Haemoglobin molecules come in different combinations of these globin chains: 98% of adult haemoglobin (Hb A) consists of four globin chains, with two identical alpha and beta chains ($\alpha_2\beta_2$). About 2% of adult haemoglobin (Hb A$_2$) consists of two alpha chains and two delta chains ($\alpha_2\delta_2$). Less than 1% of our haemoglobin (Hb F) is made of two beta chains and two gamma chains ($\beta_2\gamma_2$). Early embryos and foetal stages are marked by synthesis of all haemoglobins, other than the β chains.

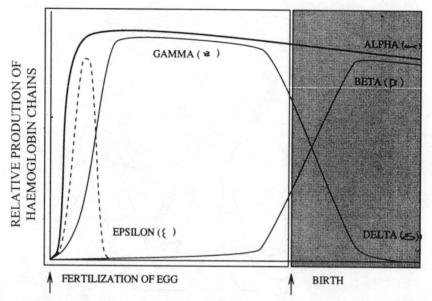

Synthesis of human haemoglobin (normal) at different stages of development. Of the four sub-units of the globin, two of them are always α throughout the life of an individual. ε and γ globin sub-units are present in the early foetal stages. β sub-units start replacing γ chains, just before the birth of a baby, with adult haemoglobin having both α and β. δ chain synthesis is negligible.

Alpha and beta thalassaemias are caused by either the absence or the presence of abnormal α or β sub-units in the RBCs. β-thalassaemia is more common than α-thalassaemia, which is often fatal in the foetal stages. The adult human population carries several changes in the β-chain of the globin molecule, which show up as different forms of β-thalassaemias. No wonder, the β-thalassaemias constitute a major health problem today in several countries of the world.

There are two types of β-thalassaemias— major and minor. A child who is thalassaemic major is highly anaemic, with an enlarged liver and a propensity to infections, and dies early in childhood. Here the gene leading to the disease is inherited from both parents. Thalassaemia minor, with mild anaemia, occurs when the gene is inherited from only one parent. Such individuals are carriers, capable of passing on the gene, and hence the disease, to their children.

The prevalence of thalassaemia is especially widespread in India, south-east Asia and several Mediterranean countries. Other blood disorders, involving the haemoglobin molecule, such as sickle cell disease and other haemoglobinopathies, are rampant among the tribals in India, in the African countries and among the blacks in the USA.

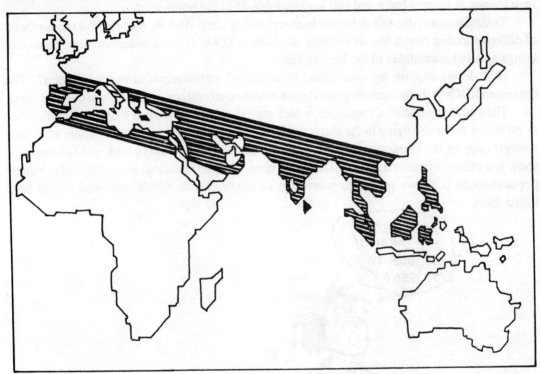

The prevalence of β-thalassaemia and other inherited blood diseases in the world. Large tracts of the equator and tropics are affected. Interestingly, malaria is also prevalent in these areas.

In India, about 3 to 4 per cent of our general population carries the gene for β-thalassaemia. These are the carriers, who could transmit the trait to their offspring. The incidence of this type of thalassaemia (minor) is alarmingly high among certain caste groups where there is considerable inbreeding, reaching as high as 10–15 per cent in several communities.

Community-wise occurrence of thalassaemia trait

Sindhi (10.2%); Lohana (7.0%); Khoja (6.8%); Bhanushali (6.9%); Sikh (5.08%); Baudth (4.1%); Bengali (4%).

Haematologists at the Indian Council of Medical Research at New Delhi and at Mumbai predict that an estimated 25 million Indians carry the genes for a variety of blood—haemoglobin—abnormalities (haemoglobinopathies), including thalassaemia. Further, they reveal that 8,000 to 10,000 infants with a severe blood disorder are born every year. This imposes a heavy load of genetic diseases on the population. Some other blood disorders involving the haemoglobin protein that are prevalent in India are: (i) haemoglobin-S mainly present in tribal and schedule caste groups, the frequency of carriers being up to 40% in some groups; (ii) haemoglobin-E is seen mainly in eastern India and (iii) haemoglobin-D in the north-west.

Thalassaemia is due to a defective haemoglobin protein. And the message for the synthesis of different globin chains lies in a family of genes in DNA. It is the defect in these genes that brings about abnormalities in the haemoglobin.

How do we explain the widespread incidence of thalassaemia in our population? The depressed RBCs in thalassaemics gives them a selective advantage, but in certain environments.

The malarial parasite, a protozoan, which reproduces in the normal RBCs, has difficulties in surviving and multiplying in the abnormal RBCs of a thalassaemic. Hence, people who carry a single copy of the thalassaemic gene—carriers (thalassaemia minors with mild anaemia)—show less effects of malaria. The thalassaemic gene-change (mutation) is maintained in human populations as it allows people to survive in an environment, which otherwise would have killed them.

'Is this exhaustion due to my tight jeans? Or is it my genes?'

Environment-gene interactions are highly complex. A lethal gene TODAY might provide protection under changed environmental conditions TOMORROW.

Sickle-cell anaemia where the individuals have sickle-shaped RBCs, is widely prevalent among certain Indian and African tribes and among blacks in the USA. In the West, there is no selective advantage in possessing the sickled RBCs, as malarial conditions are mostly absent. Nor is there a strong selection against the deleterious gene. Hence the genes persist in the population. Indeed, the high prevalence of several blood abnormalities, including thalassaemia, is a classic example of the effect of environment on the frequency of the occurrence of a specific gene in a population.

Though the focus has been on thalassaemia here, the situation is similar for other inherited diseases, too. A changed gene leads to an altered protein, which, in turn, causes a disease.

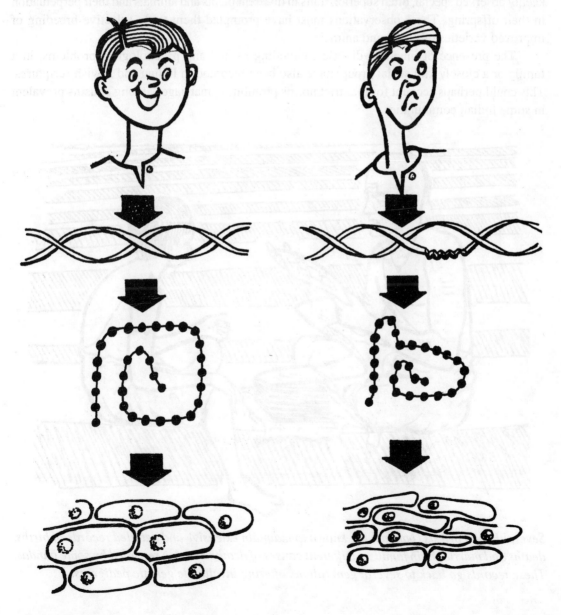

3 HOW DID IT ALL START?
Some historical ideas and the dawn of a new era

How and when did we start to build an understanding about inherited diseases? Several ancient cultures, including the Indus Valley Civilization, have left ample proof that our early ancestors keenly observed special, often superior, traits in different plants and animals and their perpetuation in their offspring. These observations must have prompted them to do selective breeding of improved varieties of plants and animals.

The presence of inherited diseases, including mental and psychological problems, in a family or a closely knit social group, have also been recorded in Indian and Jewish scriptures. This could perhaps account for the strict norms prohibiting marriages among cousins prevalent in some Indian communities.

Several Hindu priests (pandas) are known to maintain extensive and detailed records of births, deaths and marriages in families. Different causes of deaths are also recorded by some pandas. These records go back to several generations offering invaluable genetic pedigrees.

In 1865, Gregor Mendel put forward a scientific explanation for the manner in which different traits are inherited. He carried out experiments on garden pea plants. His observations are today known as the Laws of Inheritance and have rightly earned him the title of 'Father of Genetics'. He was much ahead of his time, for it was only in 1900 that his observations were validated by De Vries, Correns and Tshermack, working independently in different countries.

Mendel had artificially pollinated pea plants having different traits, and observed the precise manner in which these traits were inherited in subsequent generations. Mendel's findings are indeed brilliant, based on deep thinking, reasoning, meticulous planning of experiments and keen observations. At a time when nothing was known about meiotic cycle, chromosomes, or genes, Mendel laid the groundwork of a discipline, which is known to us as genetics today. In 1909, a Danish biologist, Wilhelm Johannsen, coined the word 'genes' from the Greek word meaning, 'giving birth to', for the invisible hereditary units carried by chromosomes.

Mendel assigned symbols for different traits : T and t for tall and short heights of pea plants.

Mendel's chosen traits were located on different chromosomes. Otherwise, he would not have put forward his laws so clearly.

What did Mendel say? How are his laws understood today? Let's understand them in the context of thalassaemia.

(1) Paired 'factors' govern all traits in an organism. It is these factors which are passed on from parents to offspring. Each parent contributes one copy of the factor to the child.

[The paired factors (genes) involved in the synthesis of the β globin protein are: HbA/HbA (healthy person synthesizing normal β globin chains), or HbA/Hba (heterozygous carrier, may or may not be mildly affected), or Hba/Hba (highly affected). HbA and Hba genes are donated by either the mother or the father.]

(2) These factors (*genes*) can exist in more than one form, called 'alleles'. The alleles are produced by changes—mutations—in the DNA of the wild type (or standard gene). There can be as many different alleles, as there can be chemical changes possible in a gene.

[The gene coding for the β chain of the globin is a good example of multiple alleles. More than 100 alleles of this gene are known, the most common being: HbA/Hba, HbS/Hbs, HbE/Hbe, and HbC/Hbc. HbA refers to the normal allele and Hba and others represent the changed alleles.]

(3) The law of segregation states that the two alleles of a gene separate during gamete formation. So each gamete receives at random either of the two alleles.

[There are several alleles of the gene involved in the synthesis of the β sub-unit. For instance, gene HbA/HbA codes for the normal chain and Hba/Hba codes for a changed chain. The genetic constitution could also be HbA/Hba. During germ cell formation, the paired genes (HbA/HbA, Hba/Hba, or HbA/Hba) separate out, to yield sperms and ova with either HbA or Hba.]

(4) These factors may or may not be expressed in an organism and, accordingly, Mendel called them as 'dominant' or 'recessive' traits. But they never disappear. His experiments categorically showed that the identity of these units (alleles) is not lost and they may reappear in subsequent generations, i.e., when two recessive alleles come together in the same individual. In other words, Mendel showed that there was no blending of inheritance.

[Strictly speaking, the haemoglobin gene is an inappropriate example to explain the dominant gene effect. But, the effect of the thalassaemic gene(s) explains the Mendelian ideas. Individuals with the genotype HbA/HbA synthesize the normal β-chain. HbA/Hba persons are also normal as the HbA gene synthesizes the normal globin chains masking the ill-effects of Hba. But these are the carriers. Hba/Hba represents the recessive condition and the individuals are diseased.]

(5) Phenotype is the organism's visible appearance. Genotype stands for its genetic make-up.

[In thalassaemic persons, weak physique, combined with anaemia, represent the phenotype. The genotypes would be: HbA/HbA or HbA/Hba (normal), Hba/Hba or HbS/HbS (diseased).]

(6) Mendel was not aware of the existence of chromosomes or meiosis. Still, his second law, the Law of Independent Assortment with respect to inheritance of two or more genes revealed an uncanny understanding of the mechanism of inheritance and has stood the test of time. Hence now we can say: genes located on different chromosomes assort independently (Law of Independent Assortment) of each other into different gametes.

[The normal globin protein of two types of polypeptide chains—α and β—is synthesized by two genes located on chromosomes 16 and 11, respectively. These chromosomes which are in

homologous pairs (and, in turn, the genes located on them) assort independent of each other during the formation of gametes.]

The law of segregation and the randomness of fertilization allow us to make predictions about the likely genotypes and phenotypes among the offspring after any mating.

Mendel's Laws are no more abstract or hypothetical. They have a strong physical basis, with DNA and its chunk of genes, providing new and amazing insights about the manner in which we inherit different characters, including some diseases. The laws of inheritance as put forward by Mendel are applicable to all sexually reproducing organisms, from the protozoans to human beings. However, the biochemical and genetical complexity in humans has revealed several extensions /deviations of these laws as reflected in multi-gene (polygenic), incomplete dominance and co-dominance inheritance.

Several human traits, such as attached earlobes, hair on the ears, pitted ears, hair on middle joints of fingers, extra toes or fingers, baldness, pattern baldness, rolled tongue, the ability to taste the bitterness of a chemical, phenylthiocarbamide (PTC), etc., are all controlled by single genes. Their inheritance follows Mendel's laws.

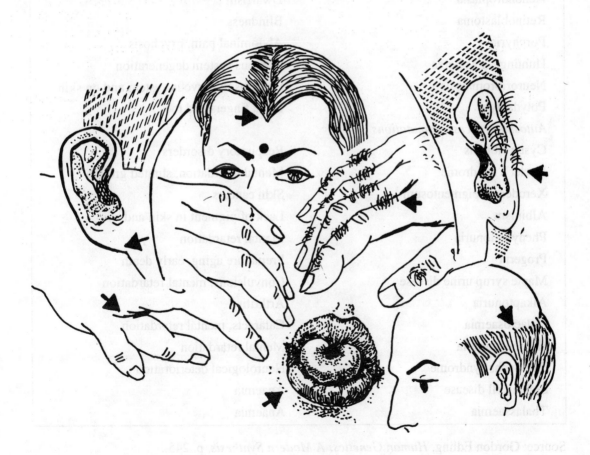

Some sex-linked diseases

Disease	Major effects
Lesch-Nyhan disease	Mental retardation
Haemophilia	Failure of blood to clot, bleeding
Duchenne muscular dystrophy	Progressive muscular weakness
Agammaglobulinemia	Defective immune system
Testicular feminizing syndrome	Sterility, lack of male organs

Some inherited diseases and disorders

Disease	Major effects
Autosomal dominant mutations	
Achondroplasia	Dwarfism
Retinoblastoma	Blindness
Porphyria	Abdominal pain, psychosis
Huntington's chorea	Nervous system degeneration
Neurofibromatosis	Growths in nervous system and on skin
Polydactyly	Extra fingers and toes
Autosomal recessive mutations	
Cystic fibrosis	Respiratory disorders
Hurler's syndrome	Mental retardation, stunted growth
Xeroderma pigmentosum	Skin cancers
Albinism	Lack of pigment in skin and eye
Phenylketonuria	Mental retardation
Progeria	Premature aging, early death
Maple syrup urine disease	Convulsions, mental retardation
Alkaptonuria	Arthritis
Galactosaemia	Cataracts, mental retardation
Homocystinuria	Mental retardation
Tay-Sachs syndrome	Neurological deterioration
Sickle cell disease	Anaemia
Thalassaemia	Anaemia

Source: Gordon Edling, *Human Genetics: A Modern Synthesis,* p. 245.

By the early 1900s, scientists immediately realized the parallels between Mendel's factors and the behaviour of chromosomes during cell division. And the two were synthesized together as the chromosome theory of inheritance according to which:

(1) there are two copies of each gene and two copies of chromosomes in all body cells;

(2) members of an allele pair and homologous chromosomes both segregate into different gametes during gamete formation;

(3) genes for different traits and non-homologous chromosomes both assort independently with the formation of germ cells.

Soon it was also realized that the sex of an organism is an inherited trait and special chromosomes, the sex chromosomes, are involved in determination of sex in different organisms. Genes located on the sex chromosomes would have a pattern of inheritance. And these genes soon came to be called sex-linked genes in humans. For instance, today more than 100 genes have been identified and located on the X chromosome, and cover such varied traits as colour blindness, Xg^a blood groups, and genes controlling the nervous, muscular and circulatory systems.

In 1865, Francis Galton introduced biometric genetics, bringing quantitative analysis to human genetics, and thus laid the groundwork for population genetics. He developed several statistical concepts in biology and was the first one to make use of twins for comparative studies. He also studied and classified fingerprints. Measurement was his obsession and he even tried to represent complex human features in terms of numbers and formulae.

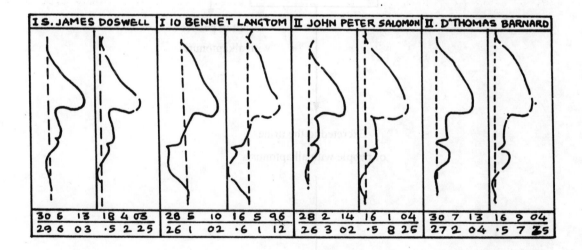

Galton used mathematical formulae to draw profiles and shapes of noses (right). Profiles from the actual pictures are on the left. Note the near similarities between the two.

Galton is considered to have given a fillip to eugenic (=well born) ideas, i.e., the science of improving people by controlled breeding. His important contributions to quantitative genetics are often overshadowed by these views. He is discredited with starting the infamous eugenic movement, some time in 1883, when he published his book, *Inquiries into Human Faculty*. Unfortunately, genetic principles were (and are) often misinterpreted due to deep rooted prejudices to justify discrimination and cruelty. Though largely discarded, certain eugenic ideas in modified forms persist in some quarters. Perhaps several eugenic ideas tainted with class prejudices discouraged serious work in human genetics in the early 20th century.

In 1902, Sir Archibald Garrod, often called the Father of Human Genetics, published a path breaking paper, *'The incidence of alkaptonuria—a study in chemical individuality'*. It was earlier observed that the urine of people affected with this disease darkened on exposure to air due to large amounts of alkapton (now called homogentisic acid (HA)) present in the urine. This excretion increased when the affected individuals were put on a diet rich in proteins, especially with certain amino acids, such as tyrosine and phenylalanine. Garrod gave an explanation for these observations: a specific enzyme was absent in the affected persons and it was this enzyme which normally broke down the HA to simple compounds. In the absence of this enzyme, HA accumulated in the urine leading to its darkening when exposed to oxygen.

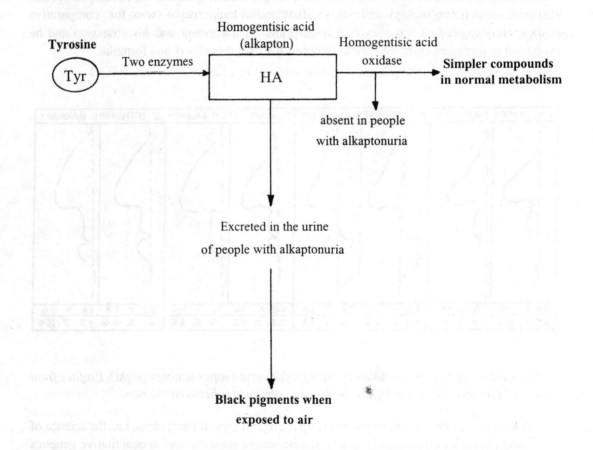

In 1909, Garrod published his classic book, *Inborn Errors of Metabolism,* bringing out strong linkages between genetic diseases and enzyme deficiencies. Thus, he established that each individual was endowed with a unique biochemistry and that diseases were '... merely extreme examples of variations of chemical behaviour which are probably everywhere present in minor degrees and that just as no two individuals of a species are absolutely identical in bodily structure neither are their chemical processes carried out on exactly the same lines.'

This is the first biochemical proof of the relationship between a disease and the presence or absence of an enzyme protein. Garrod, along with William Bateson, brilliantly connected his observations to genetics. He cleverly extrapolated Mendel's laws and observations about recessive characters to alkaptonuria patients and to several other biochemical abnormalities. Now we know that most of these diseases are indeed single-gene disorders, with the majority of them being quite rare. In fact, today, hundreds of metabolic diseases in humans have been catalogued, and more are added regularly to this list as cell metabolism is further understood. Garrod was also intrigued that most of the children with alkaptonuria had normal parents, and that these parents were first cousins. Thus, he remarked that there must be '...an explanation in some peculiarity of the parents, which may remain latent for generations, but has the best chance of asserting itself in the offspring of a union of two members of a family...'.

Father of Human Genetics, Archibald Garrod,
Quite a keen observer
And intuitively clever.
Alkaptonuria, pentosuria, cystinuria,
and albinism,
All mentioned first by him.
Frequencies of these high, where parents first cousins,
Hence forewarned us

not to marry in families.

Enzyme defects and the consequent metabolic disorders are mostly caused by genes. There are other categories of metabolic disorders, too, which arise due to defects in other types of proteins (hormones, receptors, transport proteins, immunoglobulins and collagens). For convenience, human metabolic disorders are grouped under these categories.

Early detection of metabolic disorders in a baby can greatly salvage a situation. Symptoms, such as acute illness, persistent vomiting, lethargy, convulsions and coma, abnormal decrease of sugar in the blood, jaundice, diarrhoea, abnormal urine or body odour, coarse faeces, hair and eye abnormalities and some odd physical features, should prompt investigations of these diseases.

Some examples of metabolic disorders

Disorder	Inheritance	Defective/ Deficient Protein	Clinical Features
AMINO ACID METABOLISM			
Maple syrup urine disease	AR	Branched chain keto-acid dehydrogenase	Sugary-smelling urine, breathing/ feeding problems, mental retardation, death
Ornithine transcarbamylase deficiency	XD?	Ornithine transcarbamylase	Irritability, mental/ physical retardation, fatal if untreated (FIU)
CARBOHYDRATE METABOLISM			
Hereditary fructose intolerance	AR	Fructose-1-phosphate aldolase	Hypoglycemia/vomiting after fructose intake; fatal if untreated
Galactosemia	AR	Galactose-1-phosphate uridyl transferase	Inability to digest milk; vomiting; jaundice, cataracts; mental retardation; FIU
NUCLEIC ACID METABOLISM			
ADA deficiency	AR	Adenosine deaminase	Immunodeficiency disease; FIU
Gout (one of many types, with various causes)	XR	Phosphoribosyl pyrophosphate synthetase	Active enzyme leading to excess of purines and uric acid; deafness
ORGANIC ACID METABOLISM			
Glutaric acidemia type I	AR	Glutaryl-CoA dehydrogenase	Brain damage, leading to uncontrolled movements; FIU
MCAD deficiency	AR	Medium chain acyl-CoA dehydrogenase	May resemble Reye syndrome; vomiting, hypoglycemia, lethargy; FIU
LIPOPROTEIN AND LIPID METABOLISM			
Familial LCAT deficiency	AR	Lecithin; cholesterol acyltransferase	Corneal opacities; anaemia, kidney problems
METAL METABOLISM			
Haemochromatosis	AR	?(Defect in iron absorption)	Excess iron damages liver, heart, pancreas, skin, joints; FIU
Menkes (steely hair) disease	XR	Possibly copper transporting ATPase (Defect in copper absorption)	Copper deficiency; greyish, broken hair, abnormal facial features, cerebral degeneration, bone and arterial rupture; FIU
Wilson disease	AR	?(Defect in copper transportation)	Excess copper; ring around iris, tremor, emotional and behavioural effects

contd...

LYSOSOMAL ENZYMES			
Hurler syndrome (Mucopolysaccharidosis I)	XR	α-L-iduronidase	Enlarged liver, spleen; skeletal deformities; coarse facial features, large tongue, hearing loss, corneal clouding, heart disease, mental retardation, death
Hunter syndrome (Mucopolysaccharidosis II)	XR	Induronate sulphate sulfatase	Severe form, coarse facial features, short stature, skeletal deformities, joint stiffness, mental retardation, death
Gaucher disease, type I (adult, or chronic form)	AR	Glucocerebrosidase	Variable, enlarged spleen, bone defects and fractures, arthritis
PEROXISOMAL ENZYMES			
Adrenoleukodystrophy	XR	Possibly ALD membrane transport protein	Variable; excess of very long-chain fatty acids; adrenal insufficiency. Childhood form includes dementia, seizures, paralysis, loss of speech, deafness and blindness; FIU
Zellweger syndrome	AR	?(No peroxisomes)	High forehead and other facial features; weakness; defects of eyes, brain, liver, kidneys, heart; FIU
HORMONES			
X-linked ichthyosis	XR	Steroid sulfatase	Dry, scaly 'fish-skin', mild corneal opacity
Goitrous cretinism (one of many types of hypothyroidism)	AR	Iodotyrosine dehalogenase	Dwarfism, mental retardation, goitre, coarse skin and facial features
BLOOD PROTEINS			
von Willebrand disease, type I	AD	von Willebrand factor	Varies greatly; bruising, bleeding from gums, cuts and gastrointestinal tract; skin haematomas; heavy menstrual bleeding
Haemophilia A	XR	Factor VIII, sub-unit a	Spontaneous bleeding, especially into large joints and muscles; haematomas, chronic arthritis; can be FIU
CONNECTIVE TISSUE PROTEINS			
Osteogenesis imperfecta, type I	AD	Procollagen type I	Osteoporosis, bone fractures, blue sclera (outer layer of eyeball), hearing loss
Ehlers-Danlos syndrome, type IV	AD	Collagen, type III	Fragile skin; bowel, arterial, uterine rupture
Marfan syndrome	AD	Fibrillin	Long fingers and limbs; chest deformity; loose joints; curved spine; displaced lens, heart problems

AR = autosomal recessive   XR = X-linked recessive
XD = X-linked dominant   AD = autosomal dominant

For a long time there were no clues about the unit of inheritance and its functions. In 1906, a brilliant American scientist, Thomas H. Morgan, aided by his equally committed and brilliant colleagues, many of whom were his students, launched an effort in this direction. They worked on the common fruit fly, *Drosophila*, which are easy to raise and breed fast, and hence make good experimental material. And Morgan was an expert experimentalist. Morgan, Calvin B. Bridges, Alfred Sturtevant and H. J. Muller, affectionately called as The Fly Room Boys, carried out several clever experiments. They worked out several details of basic genetics: sex determination, sex-linked genes, crossing over and gene mapping, chromosomal aberrations and non-disjunction.

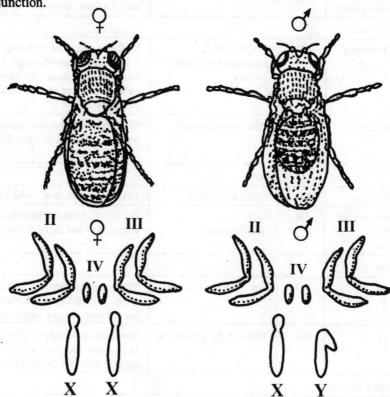

Fruit flies have four chromosome pairs, three pairs are identical—autosomes—and one pair is the sex chromosomes (XX in female and XY in male).

Morgan and his colleagues noted the usual and not so usual characteristics of Drosophila, such as the white eyed flies versus the normal red eyed ones. Mating (crossing) of the normal red-eyed flies with those with white eyes, led them to state that a mutation—a change in the genetic material—had altered the colour of the eye. Further, crosses of the flies led them to conclude that the eye colour gene is located on the X chromosome in the fly. In other words, the eye colour is sex-linked or X-linked. This was the first location of a gene on a specific chromosome.

A lot has been written about The Fly Room and the scientific atmosphere created therein by Morgan. It was haphazard, crowded and unconventional. Mashed bananas were an important ingredient of the culture medium, and the smell in the room was often nauseating, inviting the wrath of colleagues in adjacent laboratories. Sloppy and careless in his personal habits, he was often mistaken for a sweeper. But the atmosphere in the Fly Room was stimulating.

T. H. Morgan and an artist's impression of the Fly Room.

As described by Sturtevant in 1959: 'Each carried on his own experiments, but each knew exactly what the others were doing, and each new result was freely discussed. There was little attention paid to priority or to the source of new ideas or new interpretations. What mattered was to get ahead with the work....There can have been few times and places in scientific laboratories with such an atmosphere of excitement....This was due in large part to Morgan's own attitude, compounded by enthusiasm combined with a strong critical sense, generosity, open-mindedness, and a remarkable sense of humour.'

In 1933, Morgan received the Nobel Prize in Physiology or Medicine. But he did not attend the ceremony. Perhaps he was too busy. Or, the idea of a formal dress-coat put him off.

Other experiments led Morgan to important conclusions:

(i) Chromosomes are the bearers of units of inheritance.

(ii) Each chromosome carries several genes located along its length.

(iii) Several exceptional progeny of the flies were often obtained in certain crosses. These were due to abnormalities in the number of chromosomes, attributed to the phenomenon of non-disjunction, when two chromosomes failed to separate (disjoin) during meiosis. The resultant abnormal eggs or sperms led to the formation of the not-so-normal flies.

Some other logical questions which Morgan posed and answered were: How were the genes located on the same chromosome inherited? Do they separate out during meiosis (assort independently) or are they inherited together? A series of experiments soon revealed that :

(a) Genes located on the same chromosome tended to be inherited together and gave rise to the same combination of parental traits in the progenies. This tendency of the genes to remain together along the length of the chromosome was termed as linkage. **BUT**...

(b) ...exceptions were also observed with new combinations of characters in the offspring due to genetic recombination. *Recombination occurs when homologous chromosomes break at certain points and rejoin with exchange of genetic material. This swapping of genetic material gives rise to gametes with new genetic composition. This process is associated with specialized chromosomal structures—chiasmata. The number of chiasmata per chromosome varies widely.*

Recombination of genes has far reaching implications in evolution and inheritance of genes. The process throws up new combination of genes and also allows geneticists to locate genes (map them) with respect to each other along a chromosome. Genetic mapping involves two steps: identifying the chromosome on which a given gene lies, and the position of the gene on this chromosome in relation to other genes. The percentage recombination between pairs of genes lying on a chromosome depends on the distance between the two genes and this percentage is used as a measure of distance between them. Putting it simply, consider two gene loci, A and B on a given chromosome. As long as there is no chiasmata formation between these two genes, they will be passed on to the offspring together. However, if crossing over takes place between

POSSIBLE GENE COMBINATIONS
IN GAMETES

Outcome of a hypothetical crossing over involving a pair of genes. Note the possible variety of gametes with respect to these genes.

them, recombinant chromosomes will be formed. The parental traits (A and B) will now be separated, and the offspring will have either A or B. The number of recombinants will be larger if the genes are far apart. The frequency of recombinants can hence be used as an indirect measure of the distance between two gene loci. For instance, if the percentage of recombinants in a cross is 2, the concerned genes are said to be two map units apart.

In 1911, E. B. Wilson located the first human gene, the gene for colour blindness, on the X chromosome by following its pattern of inheritance. But developments in human genetics were slow.

The period between 1925s and 1950s is marked by several important findings in microbial genetics, all geared to find out the nature of genetic material. In 1928, Frederick Griffith demonstrated that a specific chemical molecule had the property of changing or transforming certain traits in bacteria. This line of work was diligently followed by Oswald Avery and colleagues. After 16 long years of research they demonstrated in 1944 that this transforming substance was deoxyribonucleic acid (DNA).

In 1946, Joshua Lederberg and Edward Tatum further strengthened the idea about DNA as the genetic material. Working with the common gut bacterium, *Escherichia coli*, they observed that some bacteria combined in them the characters of both the parents. They ascribed this combination of characters to exchange of DNA due to cell-to-cell contact, followed by recombination.

Who would have thought that bacteria and viruses would provide answers to several important genetic concepts!

The work of these and several other scientists established the distinction between the genetic material, DNA, and the products of its expression, proteins. This view became an implicit basis for subsequent studies.

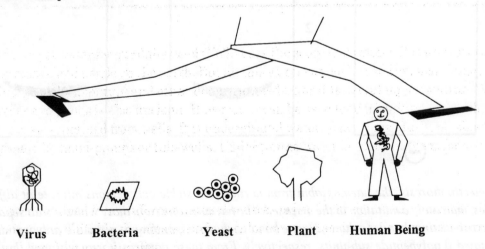

Virus   **Bacteria**   **Yeast**   **Plant**   **Human Being**

Extrapolation of results from the humble microbes to humans took time!

Meanwhile, George Beadle and Edward Tatum linked the genes with enzymes, which are a category of proteins acting as biological catalysts, and put forward the *One-gene-one-enzyme* hypothesis. They demonstrated that a genetic change caused an alteration in the structure of the enzyme and hence its activities. They worked with several strains of the pink and fuzzy bread fungus, *Neurospora*. They obtained nutritional mutants and subjected them to rigorous genetic analysis. Their work revealed:

1. 'All biochemical processes in all organisms are under genetic control.'
2. 'These overall biochemical processes are resolvable into a series of individual stepwise reactions.'

$$A \longrightarrow B \longrightarrow C \longrightarrow D$$

3. 'Each single reaction is controlled in a primary fashion by a single gene, or in other terms, in every case a 1:1 correspondence of gene and biochemical reaction exists, such that ...'
4. 'Mutations of a single gene result only in an alteration in the ability of the cell to carry out a single primary chemical reaction.'

'...The underlying hypothesis is that each gene controls the replication, function and specificity of a particular enzyme.'

Here, it is worthwhile to recall the genius of Garrod who predicted a similar link between the genetic material and its product.

The concept of one-gene-one-enzyme had limitations and was soon modified to one-gene-one-polypeptide hypothesis. This hypothesis, which is a central idea to the entire field of genetics,

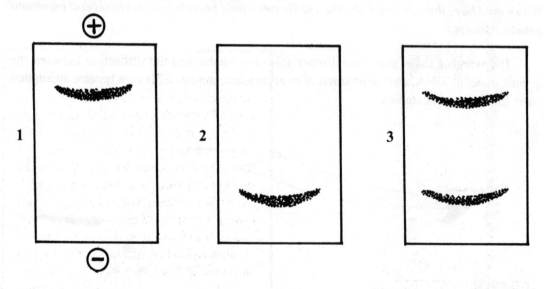

In electrophoresis, the haemoglobin from normal and sickle-cell persons migrate at different rates, indicating a mutation in the diseased. The haemoglobin from those with the trait separates into two distinct bands (column 3), one band each corresponding to normal β polypeptide and mutated β polypeptide sub-units, respectively. From these patterns it was reasoned that each gene contains information for making only one polypeptide of a protein.

was based on simple and elegant experiments by Linus Pauling and Harvey Itano in the mid-1940s. They were intrigued by the disease, sickle-cell anaemia. They reasoned that it should be possible to detect a change in the haemoglobin molecule of the sick. And they used the powerful electrophoretic technique where proteins with different charges are distinguished by their rate of migration in an electric field. Indeed, the migration rates of the haemoglobin molecule in the normal and diseased individuals were different.

Much later, in 1957, V. M. Ingram, determined the exact amino acid sequences of the two types of haemoglobins, with β chain made up of 146 amino acids, with glutamic acid in the sixth position. In the diseased, valine replaces glutamic acid at the sixth position. Just one amino acid change, and it is enough to distort the shape of the haemoglobin molecule.

Now it was the turn of viruses to be probed by biologists. The brilliant experiments by A. Hershey and M. Chase in 1952 were of decisive nature in establishing that DNA is the genetic material. They used a special type of virus, bacteriophage T2, which infects the bacteria, *E. coli*. The life cycle of these viruses is simple. First, they cling to the bacterial surface and inject their genetic material into the bacterial cell, leaving behind their protein coats. In less than 20 minutes, the bacterial cell bursts open to release more of the viral particles as the viruses have multiplied in the bacteria.

Hershey and Chase exploited this property of viruses and ingeniously labelled them with different radioactive chemicals, phosphorus 32 (P^{32}) and sulphur 35 (S^{35}). While P^{32} incorporated

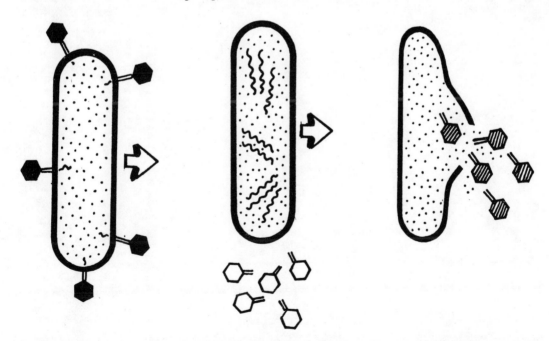

(1) A bacterial cell is infected with labelled phages. (2) Phage protein coats are left behind and the DNA is injected into the bacteria. (3) Newly synthesized phages have large quantities of P^{32} label in their DNA and little of S^{35} label.

itself into the DNA molecule as the nucleic acid is rich in phosphorus, S^{35} latched onto the protein component of the virus. Next, they detected the presence of these labels in infected bacteria. Most of the P^{32} was present in the infected bacteria, indicating the presence of DNA. The released viral progenies contained nearly 30 per cent of the original P^{32} label and less than 1per cent of the protein, as reflected in the S^{35} label.

This experiment beautifully showed that it was the original viral DNA which had entered the bacteria, as some of this DNA had become part of the progeny phages. Gradually, but surely, DNA emerged as the **UNIVERSAL** genetic material.

4 TRACING THE INHERITANCE OF DISEASES
Chromosome study and pedigree analysis

The developments in microbial genetics, unfortunately, did not have enough impact on those scientists who were studying inherited diseases among humans. They were busy studying the human chromosomes and making detailed family trees and pedigrees of individuals with genetic diseases, such as haemophilia or Alzheimer's. Human geneticists relied rather too heavily on these two approaches, which had considerable power but also great limitations.

In the 1950s, human chromosomes were being peered at closely by geneticists. They were trying to establish relationships between chromosomal abnormalities, as reflected in changes in chromosome number or structure, and certain genetic disorders. In this approach, they used special dyes and labelling techniques.

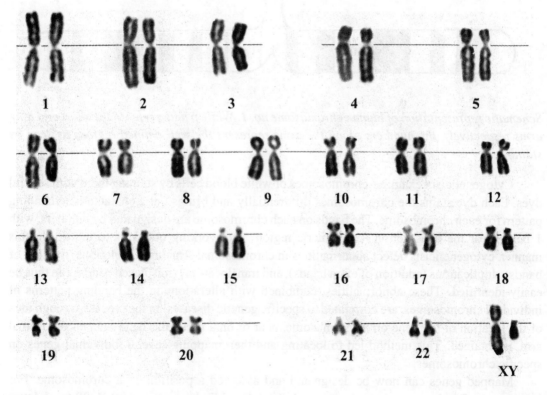

A photograph of human chromosomes (idiogram).
(Courtesy: Smt Motibai Thackersey Institute of Research in the Field of Mental Retardation, Mumbai).

The number, size, and shape of the chromosomes are specific for every organism. Microscopic study of the diploid chromosomes in humans—their structure and number—can give valuable clues regarding the presence or absence of genetic diseases.

Human chromosomes are numbered and identified according to a universally accepted nomenclature adopted in 1971. Each chromosome has two arms, connected by a centromeric region. The long arm of each chromosome is designated as 'q' and the short arm as 'p'.

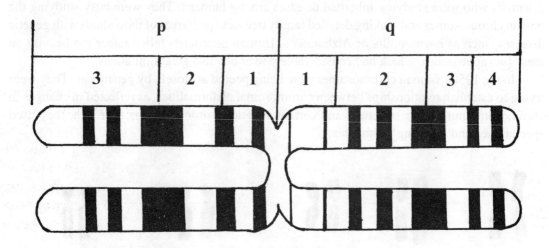

Schematic representation of human chromosome no. 1. While p and q refer to the short and long arms respectively, the numbers along the arms represent the well defined regions as seen by staining.

Cytogeneticists examine chromosomes of white blood cells by staining them with several dyes. Each dye stains the chromosomes differentially and brings out a characteristic banding pattern for each chromosome. The bands on each chromosome are designated by numbers, with 1 being near the constriction (centromeric region) and working outwards to the tip. In this manner, cytogeneticists detect abnormalities in chromosomal structures. Deletions (missing of bands), duplications (addition of extra bands), and translocations (misplaced bands) can thus be easily identified. These abnormalities, combined with alterations in the banding patterns of individual chromosomes, are correlated to specific genetic diseases. In the process, a rough idea of the location of the gene on a chromosome, near or on a particular band on a chromosomal arm, is obtained. This method led to locating and then mapping several individual genes on specific chromosomes.

Mapped genes can now be designated and assigned a position on a chromosome. For instance, the first human gene to be mapped, the colour blindness gene, is at Xq28 (each letter is to be pronounced separately). That is, this gene is located on the long arm of the X chromosome, in region 2, band 8.

What is karyotyping and when is it undertaken? During the metaphase stage of cell division, chromosomes assume a form which makes them easy to count and study. When such paired chromosomes are arranged sizewise and numbered from the largest to the smallest, it is referred to as karyotyping. Karyotyping is generally carried out in individuals who exhibit certain oddities. It is also recommended at the prenatal stage in pregnant women with an abnormal first child or repeated abortions. Several abnormalities involving the sex chromosomes also show up in karyotypes. First, one checks for variations in chromosome number. These abnormalities occur due to disturbances in meiosis during the germ cell formation in either of the parents. Often, two chromosomes of a pair do not separate out leading to a condition called non-disjunction. This leads to the formation of germ cells with abnormal—either more or less—number of chromosomes. This can happen with the sex chromosomes or with the autosomes, though autosomal fluctuations are less tolerated.

Instances of non-disjunction of sex chromosomes are far more common. If the 2 X chromosomes during egg formation in a female fail to separate out, the resulting egg could have no X chromosome or two X chromosomes. The male sex chromosomes also experience frequent non-disjunctions during sperm formation leading to sperms with either no sex chromosomes, or with chromosomal constitutions of XY, XX or YY. For reasons not yet known, the X chromosome is most frequently prone to non-disjunction.

Most of the abnormalities involving sex chromosomes are non-disjunctional and are not inherited.

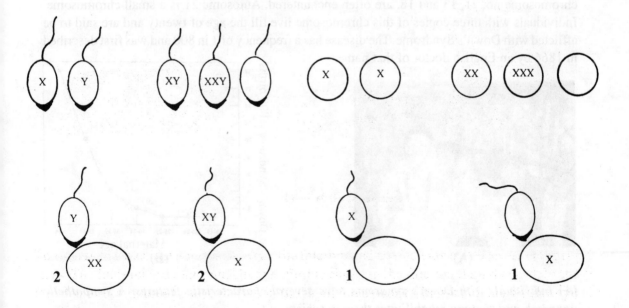

Instances of non-disjunction of sex chromosomes.

Non-disjunction of sex chromosomes can result in the following abnormalities :

(**1**) Persons affected by Turner's Syndrome (XO) have a thick neck, short stature and immature sexual characteristics.

(**2**) Persons with Klinefelter syndrome (XXY) have immature sex organs, but many have breast development.

(**3**) Metafemales (XXX) are not extra-feminine. They do not have any major physical disability, though some have learning problems. In addition, they have menstrual problems.

(**4**) Some males inherit 2Ys and 1X (XYY) chromosomes and are unusually tall. Disturbances in the number of Y chromosome does not cause severe mental or physical problems. This indicates that though the Y chromosome is important for maleness, it might not be carrying many important genes.

True hermaphroditism is a rare condition in humans, with only about a few hundreds being reported in medical literature. Technically, the term hermaphroditism is used when both the testicular and ovarian tissues are present in a single individual. And these tissues have cells with a mixture of either XX or an XY chromosomes. In contrast, those with ambiguous genitals present a still more complex problem. These persons often adapt to a sex of rearing, regardless of chromosomal constitution or development of secondary sexual characteristics.

Abnormalities in chromosome number involving the autosomes are rarely encountered in adults as foetuses abort spontaneously. Alternatively, even if such a foetus does complete its term in the uterus, and is delivered, the baby dies immediately. However, there are some exceptions. Individuals with three copies of certain chromosomes (trisomies), especially for chromosome no. 21, 13 and 18, are often encountered. Autosome 21 is a small chromosome. Individuals with three copies of this chromosome live till the age of twenty and are said to be afflicted with Down's Syndrome. The disease has a frequency of 1 in 800 and was first described in 1866 by an English doctor of that name.

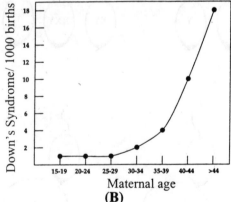

(**A**)   (**B**)

(**A**) *Individuals with Down's syndrome have several characteristic features. A sympathetic approach and training rehabilitates them in society.*

(**B**) *Down's syndrome seems to have a strong correlation with the age of the mother during pregnancy.*

Down's syndrome, like other non-disjunctional diseases, is not a typical hereditary disease: it is not passed down the family lineage as the individuals are mentally retarded (and/or sterile). However, as the germ cells carry the chromosomal abnormality, it is grouped along with other inherited diseases.

When the chromosomes 13 and 18 are triplicated (trisomy 13 and 18) as in Patau and Edwards syndrome, children are mentally handicapped and physically abnormal respectively. Such children die within a few months after their birth.

Fragile X-syndrome is yet another category of chromosomal abnormality. This is the most common familial form of mental retardation with a frequency of about 1 in 1,250 among male and 1 in 2,000 among female children. The typical characteristics are moderate to severe mental retardation, long faces and large ears. As its name suggests, one or both X chromosomes lose their fragile tips at specific positions during germ cell formation (meiosis). The available evidence indicates that the severity of the disease (penetrance) somehow increases with each generation. Recently, the gene responsible for this syndrome has been isolated and characterized, and the disease state is attributed to large tracks of triplet repeats of CGG in this gene. While normal individuals can contain upto 200 triplet units, in affected males these repeats can expand from several hundred to thousands.

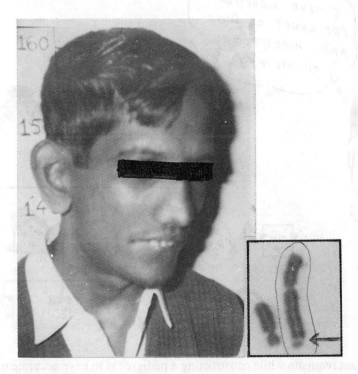

Person with fragile X chromosome (Inset*).
(Courtesy: Smt Motibai Thackersay Institute of Research in the Field of Mental Retardation, Mumbai).

Karyotypes of aborted foetuses reveal a variety of chromosomal abnormalities, including loss of one or more chromosomes (aneuploidy), extra sets of chromosomes (polyploidy), and translocational structural changes in the chromosomes with deletions and exchange of chromosomal parts. Foetuses with such abnormalities are spontaneously aborted.

Simultaneously, attempts were also made to make a family tree or a chart as extensive as possible, including uncles, aunts and their spouses and children, running into several generations, and check for the presence or absence of a genetic disease in them. This method of studying genetic diseases is called pedigree analysis. The manner in which the disease appears or disappears among different family members, over several generations could tell whether the trait concerned (perhaps a disease, too) is due to a dominant or recessive gene, or the gene is located on a sex or autosomal chromosome. Most important, pedigrees enable us to calculate the probabilities of whether a particular trait would be inherited or not in a given progeny. In other words, following a disease in a pedigree allows one to predict and calculate the chances of inheritance of a particular disease (gene).

An important requisite while constructing a pedigree is to have accurate medical histories of all family members, spanning as many generations as possible. (And not to forget the illegitimate births, adoptions, abortions, artificial inseminations, and hazy memories.) Why not start tracing your family tree? It can be an exciting, perhaps also an unnerving project.

The rules for analysing a pedigree follow Mendel's principles. Some universally recognized symbols are used by geneticists to construct pedigrees, and then trees are made by genetic counsellors.

Males are represented by a square, females by a circle. Parents are joined by a horizontal line.	
Siblings are listed in chronological order of birth.	
Fraternal (dizygotic) twins are joined by a partial triangle.	
Identical (monozygotic) twins are joined by a closed triangle.	
Affected persons (those having the disease or trait) are indicated by filled-in circles.	
Heterozygous carriers of recessive autosomal traits are partly filled in.	
Female carriers of sex-linked traits are indicated by a circle with a filled-in dot.	

Symbols used by geneticists to construct pedigrees.

Pedigree analysis essentially gives information about genetic traits determined by single genes. These traits may be due to dominant or recessive genes, located on autosomes or sex chromosomes. *McKusick's Classic Genetic Catalogue,* which is regularly updated, catalogues 'proven' traits. While a large number of traits are autosomal dominants, the mode of inheritance of several others is not yet established.

There are several traits (and diseases) with a genetic base which are difficult to follow in a pedigree. Here, more than one gene is implicated. Behavioural disorders like schizophrenia and complex diseases, including cancer, are difficult to study in a pedigree.

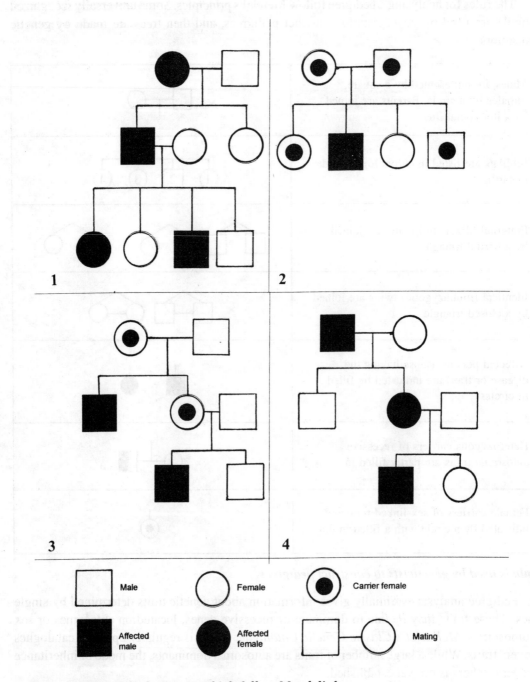

Different patterns of inheritance which follow Mendel's laws.
(1) Autosomal dominant inheritance; (2) Autosomal recessive inheritance; (3) X-linked recessive
inheritance; (4) X-linked dominant inheritance.

Let's discuss a few inherited diseases by looking at their pedigrees.

One of the extensive pedigrees which has been studied is that of the Royal family in Europé, with several of its members afflicted with a blood disorder, haemophilia. Those with this disease could bleed to death at the slightest scratch or wound. This is due to the absence of a blood protein, which helps in the formation of a blood clot. Pedigree analysis established that haemophilia is caused by a recessive gene on the X-chromosome, as only the male members of the family were affected.

Today, however, a few female members with haemophilia are known, as the males with better treatments survive, marry and produce children. For the females to be affected, two defective X chromosomes (XX) must be present (one each from the father and the mother). (See p. 53, pattern no. 3.)

Some relevant points

1) An affected son can have parents who have normal phenotypes.

2) For a female to have the characteristic, her father must have it.
   Her mother should also have it or be a carrier.

3) A characteristic could skip a generation or two.

Source: Mange, Elaine Johansen, and Arthur Mange: *Basic Human Genetics*.

Colour blindness (inability to distinguish between red and green colours) and Duchenne Muscular Dystrophy (DMD: a degenerative muscle condition) are other sex-linked recessive genetic diseases, affecting males. The manner in which the disease appears or disappears over generations, affecting only the males, strongly indicates an X-linked recessive disease.

Muscular dystrophies represent a group of genetic diseases where the body's skeletal muscles waste away in a gradual manner. There are several manifestations of these dystrophies, ranging from mild wasting away of muscles to near-crippling conditions, which could turn out to be fatal when one is just 20 years of age. DMD is the most crippling form affecting about 1 in 3,000 male births. It is a sex-linked genetic disease.

Young children affected with DMD are identified easily by their characteristic wobbly gaits and the manner in which they get up from a sitting posture. This is due to weak muscles in the leg and in the skeletal back regions, though the calves are characteristically thickened. There is a tendency to ignore these symptoms, though early diagnosis can greatly reduce the misery of a child. Intelligence is normal in most patients, and, in fact, several are gifted, like Anand A. Bantwal of Goa.

ANAND /6-95

This is a sample of one of the nine greeting card designs made by Anand Arun Bantwal, President, Indian Muscular Dystrophy Association (IMDA), Goa. The sale of these cards sustain the Goa Chapter. IMDA is a charitable society for the welfare of those with DMD and other allied neuro-muscular diseases. Besides spreading awareness about the disease, the Association provides counselling for the patients and families, networks with government and non-governmental organizations for the health, education, rehabilitation, recreation, vocational training and employment of the disabled. IMDA also promotes research in dystrophies and in the development of orthopaedic and prosthetic aids and services.

Anand's address: Indian Muscular Dystrophy Association (Goa)
'Anand', No. 15, Defence Colony, Alto Porvorim, Goa-403521
Tel: (0832)-217605; Fax: (0832)-232754

Becker muscular dystrophy (BMD) represents a milder version of DMD, with all the symptoms cropping up much later in life. There is no cure or effective treatment for DMDs. However, attempts can be made to somehow slow down the deformity of the joints and tendons, and increase muscle strength. This is partially achieved by the use of light weight splints and braces, and with physical and pulmonary therapy.

The inheritance pattern of the disease shows a classical X-(sex) linked pattern, with healthy mothers (who are the carriers) transmitting the disease gene to 50 per cent of their sons. In other words, 50 per cent of male children of the carrier mothers are likely to be affected while the other 50 per cent will be healthy. The mothers have one disease and one normal dystrophin gene, with the mother's normal copy compensating for the one causing the disease. The dystrophy shows up in boys as there is only one X chromosome, with no compensatory chromosome.

Advances in molecular biology and biochemistry have pin-pointed the culprit in DMD: an absence of a critical protein, dystrophin, in the muscle fibres, specifically in the membrane of the muscle cells, that causes muscle degeneration. And the gene for this protein is located on the short arm of the X chromosome. The gene coding for dystrophin is the largest gene so far discovered, occupying about 20 per cent of the short arm of the X chromosome. Its large size accounts for the many types of dystrophies. The longer the string of the DNA for a protein-product, the greater the chance that a random change, especially a mistake, will happen in that gene. While nearly two-thirds of the affected boys inherit the gene from their 'carrier' mothers, one-third of the DMD cases are due to new mutations in this large gene.

In India, different types of muscular dystrophies are prevalent, especially in small inbreeding groups. It is not unusual to see groups of people in small communities/tribes affected with a variety of dystrophies. There are records of the high incidence of DMD among children in several villages in South India.

Talking about sex-linked traits, there are also instances of X-linked dominant and Y-linked inheritance. Vitamin-D resistant rickets (hypophosphatemia) and a rare blood group—Xg (both instances of X-linked dominant)—are encountered in both heterozygous and homozygous females and again among the males, as the gene is located on the X chromosome. Not surprisingly, the X-dominant traits are twice as common in females as in males. The affected males pass on the condition to their daughters but never to their sons.

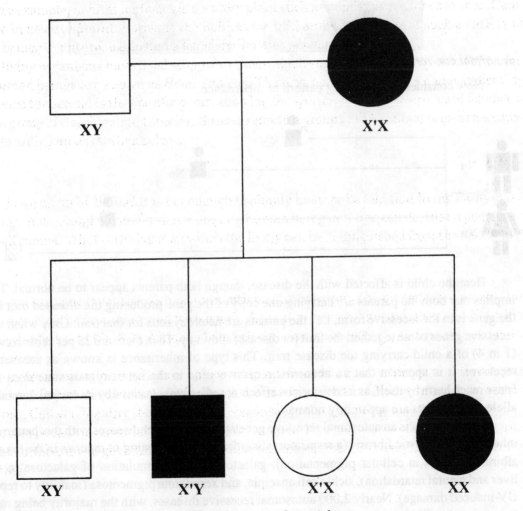

Inheritance pattern of sex-linked dominant gene, such as rickets.

Check out the hairy ears of your friends. Since the gene for the hairy ears is present on the Y chromosome, only males are affected. Try tracing out the inheritance pattern (Y-linked pattern) of this characteristic in your family, if you have hairy ears. But, oops! only in the males.

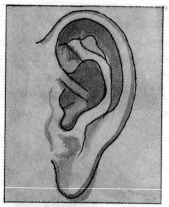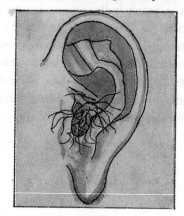

A normal ear versus a hairy one.

Now consider the following pattern of inheritance.

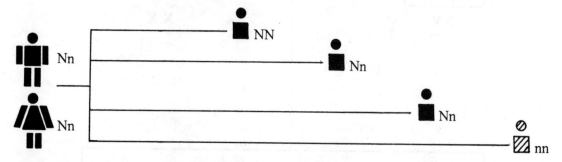

Here the child is affected with the disease, though both parents appear to be normal. This implies that both the parents are carrying one copy of the gene producing the diseased trait and the gene is in the recessive form, i.e., the parents are heterozygous for that trait. Only when two recessive genes come together, the trait (or disease) shows up. Then there are 25 per cent chances (1 in 4) of a child carrying the disease trait. This type of inheritance is known as autosomal recessive. It is apparent that an abnormal recessive gene in the heterozygous state does not cause much harm by itself, as its deleterious effects are effectively masked by its normal dominant allele. The carriers are apparently normal.

Thalassaemia is an autosomal recessive genetic disease. Other diseases with this pattern of inheritance are: cystic fibrosis (a respiratory disorder due to thickening of mucous in the lungs), albinism (defect in cellular pigmentation), galactosaemia (accumulation of galactose in the liver and mental retardation), sickle cell anaemia, and xeroderma pigmentosa (inability to repair UV-induced damage). Nearly 2,000 autosomal recessive diseases, with the majority being rare, have been described so far. Pedigrees of these diseases could be confusing. Pedigrees often digress from the predicted patterns, especially when the families are small and when the trait is absent in a generation or two. This can happen when heterozygous individuals marry outside their families, reducing the chances of mating of two carriers.

1) Most affected children have normal heterozygous parents.
2) Two affected parents will always have affected children.
3) Affected individuals who have non-carrier spouses will have normal children.
4) Both females and males are affected with equal frequency.

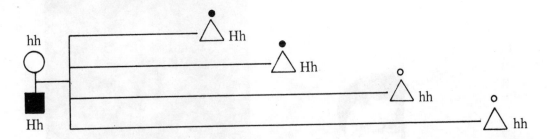

Here one parent is affected with the trait, which also shows up in the child. The parent apparently has either one or two copies of the dominant gene. Now, if an affected individual marries a normal individual, and all children are affected, then one can say that this parent carries two copies of the trait, and is homozygous dominant. On the other hand, if only 50 per cent of children are affected, then the parent has only one copy of the trait, and is heterozygous dominant (as shown above).

Autosomal dominant traits present interesting situations. If the dominant gene is harmful and reduces the life span or the reproductive ability of the individuals, such individuals are rapidly eliminated from the population. This especially happens when there are two copies (homozygous) of such genes, as in familial hypercholesterolemia (FH), characterized by high levels of cholesterol in the blood. Other autosomal dominant diseases present a more insidious face. The initial symptoms may be mild or crop up later in life, after an individual has given birth to children. This leads to an increase in the number of diseased individuals in the general population.

Some well known autosomal dominant disorders are: Marfan syndrome (affecting the skeletal system, the eye, and the cardio-vascular system), achondroplasia (dwarfism), brachyd-actyly (very short fingers), porphyria (inability to synthesize porphyrins essential for enzyme functioning, leading to a variety of clinical symptoms, ranging from mild to more lethal ones), and hypercholesterolemia and hyperlipidemia, characterized by high levels of cholesterol and lipids in the blood. If left untreated and undetected, the latter two are known to contribute to diseases of the heart and artery, and account for about 5 per cent of all heart attacks in persons under 60. Genetic counselling and prenatal diagnosis are recommended for individuals who have family histories of these diseases.

Some relevant details

1) An affected child usually has an affected parent.
2) Heterozygotes are affected.
3) Two affected parents can produce an unaffected child.
4) Unaffected parents do not have affected children.
5) Both males and females are affected with equal frequency.

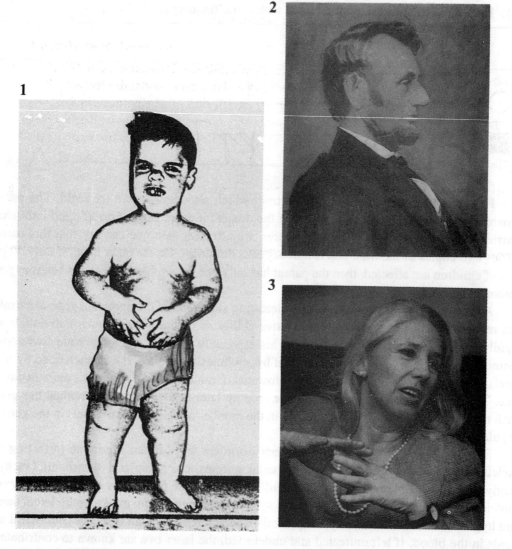

Some autosomal dominant traits: (1) A form of dwarfism, achondroplasia; (2) Abraham Lincoln: A victim of Marfan's syndrome? (3) Nancy Wexler, one of the active members of the HUGO and a potential carrier of Huntington's disease ?

Huntington's disease or Huntington's chorea is a progressive neurological disease with an autosomal dominant inheritance pattern. It holds a special fascination for human geneticists. The afflicted are considered to carry a genetic time-bomb as there is no escape from the disease. Often called, El Mal (in Spanish, literally meaning 'the bad or the sickness'), there is no way an individual can escape the disease if he/she has inherited the gene from either of the parents. The symptoms, first gradually, and later rapidly, start developing only in the early 40's or later. By then the individual concerned has unwittingly passed on the genes to his/her children, thus leaving behind a chain of individuals with the dreaded disease. Being an autosomal dominant gene, it is passed on to about 50 per cent offspring from a parent carrying the diseased gene.

This devastating neurological disorder strikes both men and women, often when they are in the prime of their life. What starts as a mere fidgeting and an excited nature, gradually develops into depression, impaired memory, twitching of the limbs or facial expressions. Clinically, there is degeneration and death of nerve cells, especially in the deep recesses of the brain. Eventually, those affected lose their faculties and become totally helpless. Once stricken, the patient (and his family too) undergoes a harrowing time, with death providing a welcome relief. Till date, there is no cure or treatment for the disease.

Artist's impression about the village in Venezuela, where Huntington's disease is most prevalent.

A wealth of genetic information exists in a village in Venezuela, South America, where hundreds of people are afflicted with Huntington's disease. These people live in isolated conditions, with the remote village having acquired global notoriety. In 1955, Americo Negrete, a Venezuelan doctor, recommended that sterilization of certain members of the village population could lead to the disease being effectively wiped out. Then the suggestion was frowned upon, but now this seems to be the only way to control the spread of Huntington's, and perhaps several such autosomal diseases, in the general population.

Tracking the gene for this disease is a gripping story involving intensive and often frustrating efforts spanning nearly 11 years. James Guesella and his colleagues at the Harvard Medical School carried out detailed molecular analysis of the blood samples obtained from affected families in Venezuela and in the West by the obsessively driven Nancy Wexler. In 1983, a paper in *Nature* announced the location of the gene on chromosome number 4, at the tip of its small arm. The gene is characterized by highly repeated nucleotide sequences. The late onset of the disease implies that the gene for Huntington's is somehow activated at a particular time. Pedigree analysis initiated in the earlier part of this century has turned out to be useful to human geneticists. Today, molecular tests combined with pedigree analysis, pinpoint the presence or absence of a diseased gene, helping to evolve strategies to prevent inherited diseases.

Karyotyping and pedigree studies were the highlights of human genetics till mid-twentieth century. However, questions concerning the structure and function of a gene, the hereditary unit, remained unanswered, though several earlier experiments on microbes had established that the hereditary information is carried in a chemical molecule, DNA. And it is this molecule which is passed on from parents to their children. Vinita and Vinod want to know whether they could ever have a normal child. To answer this question, let us next find out more about the hereditary molecule.

5 DNA, THE MOLECULE OF THE CENTURY
Its structure and functions

What is being transferred from parents to offsprings?
Factors, said Mendel. Even proteins, said some.
It is DNA, deoxyribose nucleic acid.
Chromosomes are made up of DNA and proteins.

In 1953, James Watson and Francis Crick proposed the structure of DNA. Rosalind Franklin and Maurice Wilkins also played an important role in this effort.

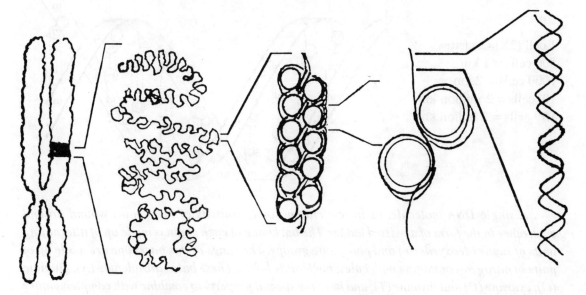

A drawing of several orders of chromatin packing in a chromosome.

The structure of DNA is indeed a biological marvel. The molecule contains the blueprint of all genetic information in a coded form. What makes my hair curly and yours straight, the eyes light or dark in colour, and the information about diseases, including thalassaemia, is all present in the DNA. Further it is this information in the DNA which is transmitted from one cell to the other dividing cells (during mitosis) and from parents to their children, via germ cells (recall meiosis). The coded information in DNA molecule is converted into varied characteristics via protein synthesis in our cells. Proteins catalyse metabolic functions and contribute to the structure of cells and communication between them. Collectively, the thousands of characters of individuals are thus determined.

DNA has a beautiful structure!

DNA can make exact copies of itself—It can Replicate.

DNA can change—It can Mutate.

DNA carries a store-house of information.

All these properties make it an ideal hereditary molecule.

DNA is a long chemical molecule organized and packed into chromosomes in a most intricate and exquisite manner. Stretch out the total DNA in a single human cell with its 46 chromosomes, and it would extend close to 2 metres. And if we join and stretch out the DNA present in all our cells (and there are about 10^{14} of them in an adult individual), the molecule could extend to the sun and back. And DNA is mere 50 trillionths of an inch across!

1 cell (2X)=2 metres
500 cells = 1 km
1000 cells = 2 km
10^9 cells = 2 million km
10^{12} cells = 2 billion km

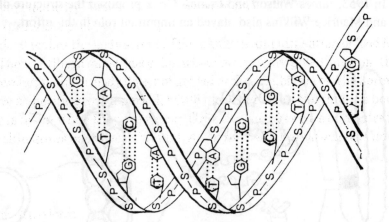

A single DNA molecule, as in one chromosome, consists of two chains wound around each other in the form of a twisted ladder. The backbone of each chain is made up of alternating units of sugar (deoxyribose) and phosphate groups. The rungs of this backbone are made up of pairs of nitrogen-containing molecules, nucleotide bases. These bases are adenine (A), guanine (G), cytosine (C) and thymine (T), and have the special property to combine with complementary bases by weak hydrogen bonds. Adenine will always pair with thymine and guanine with cytosine. This complementary nature of the bases holds the DNA molecule together and also allows it to make exact copies of itself.

When the two DNA chains separate out during replication, each chain acts as a template guiding the formation of two new chains. The free nucleotides in the cell bind to the complementary bases in the template, with the process being catalysed by an enzyme, DNA polymerase. Now the original DNA molecule has given rise to two exact copies of itself which are transmitted to its daughter cells. Normally, there are no mistakes in the replication process. The cell has evolved a whole battery of regulatory or proof-reading enzymes which roll along the length of DNA, checking and rechecking any possible errors.

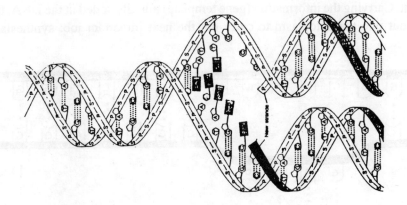

DNA replication and proof-reading enzymes.

In 1958, the central dogma of molecular biology was put forward, which has now been refined to :

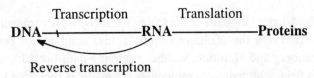

The terms transcription and translation are apt. Transcription refers to a close copying mechanism which converts the same language to a different form. Translation is putting the information in another language, wherein the DNA language based on only four letters (A, T, G, C) is translated into 20 amino acids for the formation of different proteins.

Protein synthesis takes place through a series of steps in a cell. Another type of nucleic acid, ribonucleic acids (RNAs) are involved in protein synthesis. These RNAs are also made up of linked up nucleotides (A, C, G, U), but are much smaller polymers. RNAs are single stranded molecules. The sugar in RNA is ribose instead of deoxyribose, and the base thymine is replaced by uracil. Like thymine, uracil too complements with adenine.

There are several types of RNAs, their function varying with their structure. One of the most important classes of RNAs is the messenger RNA (mRNA). The enzyme RNA polymerase helps in the synthesis of mRNA. Normally, each human mRNA molecule carries the message for only one gene, which might be several hundred to thousand nucleotides long. Only one strand of DNA is transcribed (copied) as the mRNAs, stopping and starting with the 'start' and 'stop' signals in the DNA, which are in the form of specific nucleotides. This transcription process, whereby the information carried in the DNA is copied as the RNA, is controlled by a group of small proteins which are regulatory in nature and are called transcription factors. These proteins are intricately and closely associated with enzymes involved in synthesis of the RNA and the DNA. Their structure and production is also highly influenced by internal development and several environmental signals. All this greatly affects the synthesis of mRNA, providing valuable clues as to how certain genes are expressed at certain times during development. Carrying the information (gene template) initially coded in the DNA, the mRNAs now move out into the cytoplasm to carry out the next important job: synthesis of protein polymers.

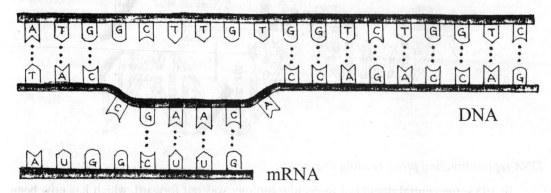

Transcription: the transfer of information from DNA to mRNA.

In translation, the 4-letter language of the DNA, which is already transcribed on the mRNA, now has to be converted into the 20-letter (amino acids) language of proteins. In the early 1960s, Marshall Nirenberg and Heinrich Matthei made the path-breaking discovery that each of the 20 amino acids from which protein molecules are made, is specified by three consecutive nucleotides—codons—in the DNA. Thus, the 4-base alphabet of the nucleic acids in DNA and

	Second base in codon				
First base in codon	**U**	**C**	**A**	**G**	Third base in codon
U	UUU } Phe UUC UUA } Leu UUG	UCU UCC UCA } Ser UCG	UAU } Tyr UAC UAA } Stop UAG	UGU } Cys UGC UGA Stop UGG Trp	U C A G
C	CUU CUC CUA } Leu CUG	CCU CCC CCA } Pro CCG	CAU } His CAC CAA } Gln CAG	CGU CGC CGA } Arg CGG	U C A G
A	AUU AUC } Ile AUA AUG (start) Met	ACU ACC ACA } Thr ACG	AAU } Asn AAC AAA } Lys AAG	AGU } Ser AGC AGA } Arg AGG	U C A G
G	GUU GUC GUA } Val GUG	GCU GCC GCA } Ala GCG	GAU } Asp GAC GAA } Glu GAG	GGU GGC GGA } Gly GGG	U C A G

The genetic code.

RNA is capable of creating 64 combinations. Since only 20 amino acids are used in the synthesis of proteins, more than one codon codes for an amino acid. Thus, 61 codons code for amino acids and the remaining three are stop codons that halt protein synthesis.

Another type of RNA—transfer RNA (tRNA)—now comes into play for synthesis of proteins in the cytoplasm of the cell. These tRNA molecules are designed to recognize a specific codon in mRNA and also carry the corresponding amino acid. Cellular machinery, including specific enzymes, catalyse the attachment of individual amino acids to each other in a sequence decoded from the mRNA with the help of tRNA. Thus a chain of different amino acids, specifying a specific protein, is built. When the protein acquires a 3-D structure, it starts to carry out different functions.

Life without proteins is just not possible. The thousands of enzymatic proteins carry out different functions, and could be involved in rapid breathing during exercises, or digesting our food, or building new tissues after an injury. The proteins also contribute to the structure of the muscle fibre, the nerve fibre, and the RBCs, among several others. The right type and amounts of proteins must be produced by a cell, the instruction of which are lodged in the DNA in our chromosomes. Any disturbances (errors) in the DNA molecule has the potential of affecting proteins.

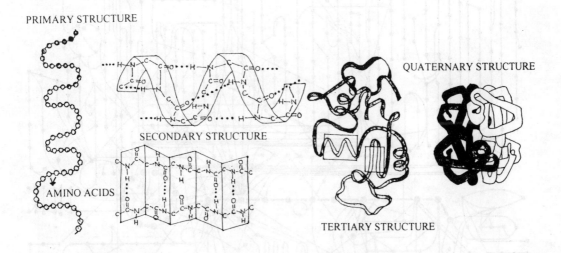

The structure of proteins.

Proteins become functional only in their 3-D form. This implies that they must bend and twist quite a bit from their original linear shape (primary structure). Interestingly, every twist and turn is precisely guided by the type of amino acids in the linear chain. While certain amino acids tend to fold inwards, others always occupy the outer shell; still others repel or attract other amino acids due to the charges carried by them. All these and several other factors determine the final shape and function of the proteins. But in the ultimate analysis, it is the DNA with its base sequences which determines the protein structure and shape.

Among thousands of genes in our cells, only a small fraction of them are expressed at any given time. Those contributing to basic cellular functions, such as respiration and general metabolism, are functional at all times in cells. And others function at certain times during the developmental process. This becomes most pronounced during differentiation and development of a foetus, when certain genes are switched off and on, to give rise to cells of the liver, skin, kidney, etc. Hence, genes are expressed in a selective manner at different stages of development.

While several genes are involved in establishing permanent features of developing cells, some come into action on a transient basis only, for example, during stress conditions. These genes are switched back to their normal activity once the stress conditions are removed.

This discussion should give some idea about the inherent complexity of living organisms, especially at the genetic level. Though diseases like thalassaemia may appear remote to a lay person from this discussion, it should be apparent that strong connections existing between genes, proteins and our lifestyles lead to inherited diseases. Understanding the molecular aspects of a cell should help one to understand the varied ways in which mutations and inherited diseases can occur.

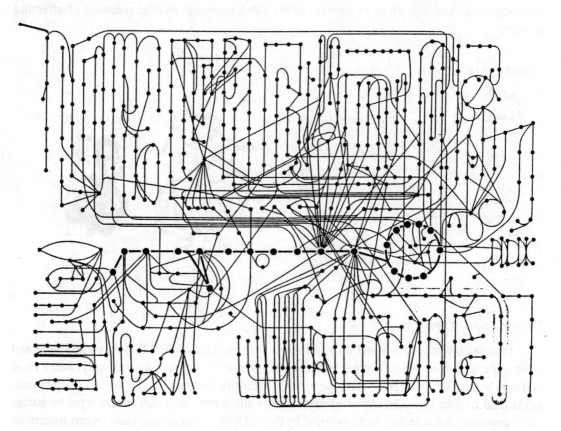

Chemical circuitry (~500 biochemical reactions) in a cell, a bird's eye view of its molecular complexity.

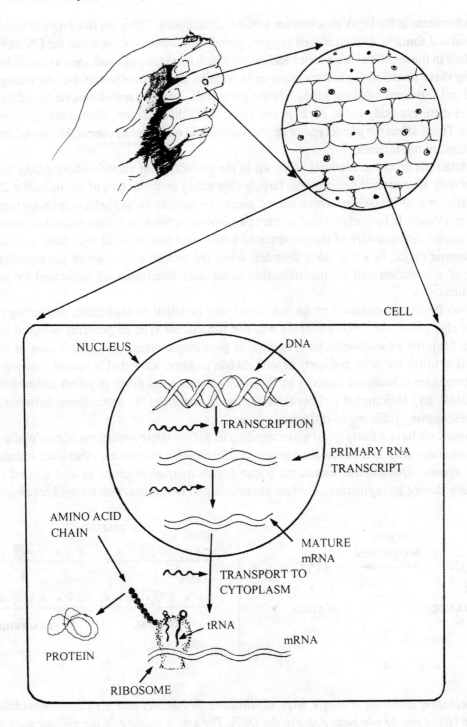

CELL

NUCLEUS

DNA

TRANSCRIPTION

PRIMARY RNA
TRANSCRIPT

AMINO ACID
CHAIN

MATURE
mRNA

TRANSPORT TO
CYTOPLASM

tRNA

mRNA

PROTEIN

RIBOSOME

A cell making proteins.

Alterations in the DNA structure are termed as mutations. There are two types of mutations: germinal and somatic. As their names suggest, germinal mutations occur when the DNA changes take place in those cells of the ovaries and testes which form the eggs and sperms. An individual carrying these mutations in his/her germ cells remains largely unaffected, but the changes are passed on to succeeding generations, via the germ cells. Somatic mutations, on the other hand, occur in different body cells, such as the cells of the liver, colon, skin, kidney, or urinary bladder. These affect the phenotype of an individual, and may lead to cancer. However, somatic mutations are not familial.

Mutations do not necessarily show up in the phenotype of an individual. Many become evident only after several generations, largely depending on the nature of the mutation. This is especially true of mutations which are of recessive nature, as in thalassaemia carriers like Vinita and Vinod. In fact, millions of us may be carriers for these and other recessive mutations. These carriers are unaware of the presence of a recessive mutation till they have a child with thalassaemia major, or some other disorder, when the two recessive genes get together. The timing of a mutation and its manifestation in an individual may be separated by several generations.

The effects of a mutation on an individual may be lethal or negligible, depending on the kind of changes in the DNA molecule and the subsequent type of proteins affected by this change. Majority of mutations have negligible phenotypic effects (like cleft chin or uncleft one, soft or brittle ear wax, and curly or straight hair). Others are lethal in nature, causing death at different ages (childhood death as in Tay-Sachs disease, and death in youth and middle age as in DMD and Huntington's chorea). Majority of mutations lie somewhere between these extremes, causing suffering to different degrees.

Today, we have a fairly good understanding as to how these mutations occur. While some occur spontaneously and hence are known as spontaneous mutations, others are induced by certain agents. Spontaneous mutations occur due to thermal motions in and around DNA, especially during its replication. Certain chemicals and radiations which could reach into the

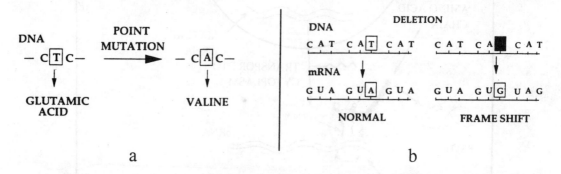

(A) A mutation involving a single base substitution. Mutations can also involve addition or deletion (B) of one or two base pairs in the DNA. The entire reading of the codons goes out of phase after this type of 'frameshift' mutation.

cells also cause these mutations. In contrast, induced mutations are caused by agents, broadly known as mutagens, which include exposure to certain chemicals, radiations and even some viruses.

Mutations involve a variety of changes in the hereditary molecule. There might be a change in just one base (base substitution or point mutation) which may or may not have a drastic effect on an individual. On the other hand, a number of bases may be altered, often leading to abnormal proteins. The effects of mutations also depends on the region of DNA affected, whether the changes are in the coding or non-coding regions.

Some mutations are beneficial and in the long run these changes establish themselves as new genes in a population. Such mutations set the evolutionary trends. Majority of inherited diseases, including thalassaemia, have originated as mutations in the population. It is these mutations that are our concern today.

Genes are lodged in two bundles in any cell: a big chunk of DNA is carried by the nucleus, and a much smaller bundle is carried by the cellular organelle, mitochondria. The human mitochondrial genome has been sequenced and there are 37 genes in it. Some of these code for RNA molecules, and others for the respiratory enzymes. Both the germ cells—sperm and egg—carry mitochondria, but during fertilization, a sperm injects only its nucleus into the egg cell. Hence, only maternal mitochondria with their DNA are transmitted from generation to generation. This transmission of mitochondrial genes gives rise to a distinctive maternal pattern of inheritance, in contrast to the Mendelian mode. Two prominent syndromes linked to mitochondrial DNA, mitochondrial myopathy and Leber's hereditary optic neuropathy, have been detected in humans. There is wasting of muscles in the former and vision is badly affected

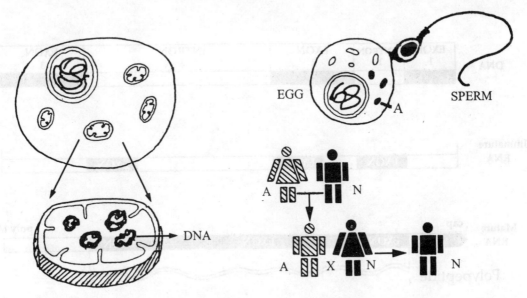

Maternal inheritance of mitochondrial genes. A: affected; N: normal.

in optic neuropathy. Mitochondrial diseases affect those tissues which have a high energy requirement, such as the heart, muscles, brain, liver and kidneys, and exhibit a marked maternal inheritance pattern. In fact, there is increasing suspicion that mitochondrial DNA might be involved in numerous and less obviously familial diseases, including Parkinson's disease, where there is death of certain brain cells.

What about the nuclear genome? How many genes do we carry? According to the estimates available in 1997, only about 3 per cent of the 3 billion individual bases of the haploid DNA in any cell actually codes for proteins. A large chunk of our genome does not get transcribed at all and we have no idea yet whether this so-called 'junk' DNA has a role to perform or not. Geneticist, Sydney Brenner, puts the number of genes in our chromosomes at just 60,000, while those with the biotechnology industry give an estimate of more than 100,000 genes!

Molecular analysis has revealed that there are similarities in the structures of several genes across different groups of organisms, with many being unique to humans. Some of these genes are small, of only a few hundred base pairs. Others are large, with several million base pairs. Interestingly, the size of a gene does not correspond to the size of the protein coded by it. This is because the code in the DNA for the protein is not continuous; it is broken by several non-coding sequences known as introns. The sequences that actually code for the proteins, which get expressed, are known as exons. Exons interspersed with introns make a gene. The entire gene gets transcribed into a mRNA molecule. However, this RNA is neatly trimmed so that all exons are intact and introns are deleted. The number of introns and exons in a gene are its unique features. The beta globin gene has three exons, while the dystrophin gene, the largest so far discovered, has 79!

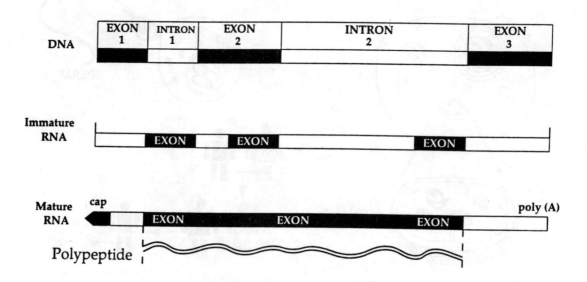

The gene coding for the β-globin chain and processing of mRNA.

Let us understand what happens at the molecular level in thalassaemia, where the globin protein in RBCs is affected. Two genes code the message for the synthesis of the globin protein (consisting of two polypeptide chains—alpha and beta). These genes are located on chromosomes 16 and 11 and exist as two separate gene clusters with other related genes. The alpha gene cluster consists of two copies of α, one of ζ (zeta), one of θ (theta), and three pseudogenes, which cannot code for functional proteins. All somatic cells have four copies of genes coding for the alpha polypeptides, two on each homologue of chromosome 16 in normal persons. Mild anaemia results if *one* of these alpha genes is non-functional, or missing. Symptoms of anaemia become acute if any two genes are missing or non-functional. The loss of three or more genes results in the death of the foetus in the womb. All these disorders due to defects in the alpha gene(s) are known as alpha thalassaemias.

The β-globin gene-cluster, too, consists of other related genes located on chromosome 11. The four other genes in this cluster are: ε (epsilon), two types of γ (gamma) and δ (delta). In the arrangement of their base sequences, all these four genes differ little from the β-gene. These globin chains are made in small amounts in the adult.

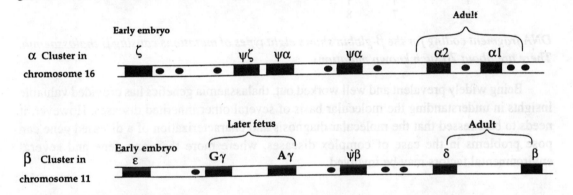

The β-gene codes for 146 amino acids. The stretch of DNA carrying the code for these amino acids consists of two introns—one large and one small—and three exons. The β-globin gene is transcribed into mRNA with all its exons and introns. This mRNA is then trimmed at proper points, the splice sites, to remove all the introns. With a bit of further modification, the mRNA moves into the cytoplasm for translation and synthesis of the beta-globin sub-units.

A variety of mutations in the β-globin gene cause β-thalassaemia.

(1) Some mutations take place in the regulatory region of the DNA. This causes the gene to be switched off and, thus, no mRNA is formed.

(2) Other mutations change a codon specifying an amino acid into a stop signal. Such 'nonsense' mutations terminate the mRNA synthesis at abnormal points.

(3) Wrong splicing sites are also made by insertion or deletion of a single base. Hence, instead of introns being removed, large chunks of useful transcripts are lost.

(4) At times, a change in a single base can have a drastic effect on the reading frame—frameshift mutation—resulting in an abnormal polypeptide.

Over the years several mutations that cause β-thalassaemia have been identified. Some of these are more common than others in a population. With molecular biology techniques, it is now possible to identify these mutations precisely. While haematologists can diagnose thalassaemia, molecular techniques can pin-point the exact changes in the DNA molecule. As in thalassaemia, all inherited diseases are caused by mutations in the DNA.

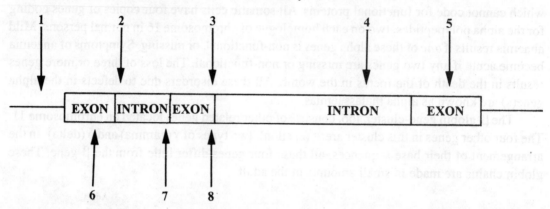

DNA fragment coding for the β-globin shows eight types of mutations causing β-thalassaemia. There are about 70 such known mutations.

Being widely prevalent and well worked out, thalassaemia genetics has provided valuable insights in understanding the molecular basis of several other inherited diseases. However, it needs to be stressed that the molecular diagnosis and characterization of a diseased gene can pose problems in the case of complex diseases, where more than one gene and several environmental factors may be involved.

those that are close to one another. The number of individuals in a generation where the genes are seen to separate from each other, or the percentage of 'recombinants', can be a clue to find the relative distance between two genes.

6 TRACKING THE GENES
DNA yields to molecular probings

How do scientists detect the gene which causes a disease? How do they know about the absence or presence of a thalassaemic gene in a foetus? How reliable are the new molecular tests? Such questions are now worrying Vinita. Two major approaches, *Forward Genetics* and *Reverse Genetics,* are being followed to pin down the gene. These strategies involve several complex steps.

'Forward genetics' is the more conventional approach. In brief, first the defective protein is extracted in large amounts and the composition and sequence of its amino acids is established. As the triplet code of bases for each of the amino acids is known, the base sequences of the messenger RNA can be worked out and thus a synthetic mRNA is made. In the next step, synthetic RNAs are labelled with an isotope so that they can be detected. This 'labelled RNA' is now used as a 'probe' to find the complementary sequences on the DNA molecule. This method works well when the defect in the protein molecule is known and when it is available in large amounts, as in thalassaemia.

In 'reverse genetics', the route to the gene is more direct. The location of the gene and its exact position on a specific chromosome is ascertained using several 'mapping techniques'. The base sequences of the gene is obtained and finally the message is deciphered to get to the protein. The gene causing Duchenne muscular dystrophy has been deciphered using this approach. This method is now being fully exploited by scientists.

To locate a gene, the first step is to identify the chromosome which carries it. As the two sex chromosomes are detected with relative ease, the genes carried by them are somewhat easier to identify. Recall the pedigree analysis, say for colour blindness (page 59). It was observed that males alone were affected. The women in these families escaped the affliction but they acted as carriers as they passed on the defect to their sons. It was obvious that wherever two X chromosomes were present, as in all the females, the defect did not surface. The defect in one X chromosome was somehow compensated by the presence of the other. As only one X chromosome was present in man, the defect in the gene could not be masked. Assuming that the colour blindness gene is carried in this pattern in affected families, the expected number of the affected individuals was estimated in each of the generations. As this estimate matched well with that actually observed, the gene for colour blindness was assigned on the X chromosome. This logic has been used extensively to 'map the genes' to the two sex chromosomes.

Once one gene is assigned to a chromosome, the position of other genes located on the same chromosome can be worked out by a simple logical analysis. If two genes are close enough, they will be inherited together and are said to be 'linked'. Genes that are far apart may not be inherited together, as they can be separated by crossing over during meiosis. These genes are more likely to be separated into two different germ cells during gamete-formation than those that are close to one another. The number of individuals in a generation where the genes are seen to separate from each other, or the percentage of 'recombinants', can be a parameter to find the relative distance between two genes.

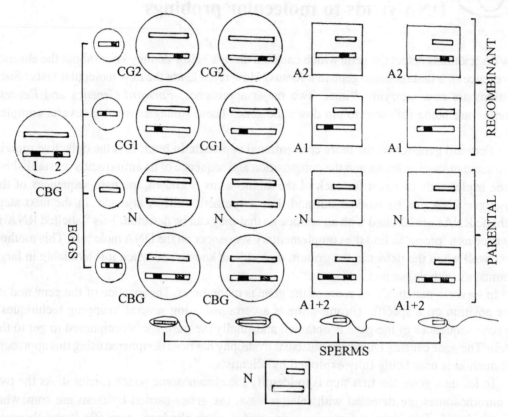

Inheritance and linkage of colour blindness and G6PD genes. (N:normal, G1:G6PD deficiency, G2:colourblind, CG1:carrier of G1, CG2:carrier of G2).

The X chromosome, for example, also carries the gene coding for an enzyme, glucose-6-phosphate dehydrogenase (G6PD). A mutation of this gene is detected as a deficiency of this enzyme. As in the case of colour blindness, women are carriers while men suffer the enzyme deficiency. In some families, both colour blindness and G6PD deficiency were seen to surface together. Some women in these families carried both the defective genes. All the children born to these women, however, were not suffering from both the problems: They were either colour blind, or had enzyme deficiency. The incidence of having only one of the defects was one in twenty, or 5 per cent. The genes in these 5 per cent individuals had separated out due to crossing over. As genes were separated five per cent of the time, they are said to be five 'map units' away.

This sort of analysis involves pedigree studies and is called 'linkage analysis'. Linkage analysis has helped in positioning relative positions of the genes. Classical linkage studies were pioneered in *Drosophila* by T.H. Morgan (see page 45). To honour his contributions, the map units are measured as 'centimorgans' (cM). The genes for colour blindness and for G6PD are said to be five centimorgans apart.

A linkage map or genetic map of a chromosome that gives the relative position of the genes can be worked out for several sex-linked diseases. From the sequence of the genes on a chromosome, the 'genetic distance' between genes can be estimated. One can compare the linkage or genetic map to a city map drawn on a piece of paper, which helps one move around the city. Similarly, with a genetic map, one can only make some estimates about the physical layout of the chromosomes. Such a map gives little information about the actual physical details of a chromosome in terms of the arrangement of bases.

The average genetic length of a human chromosome is about 140 centimorgan units, while the physical length of its DNA is about 130 million base pairs. Roughly then, one centimorgan corresponds to a million base pairs. As the frequency of crossing over is more in females than in males, the genetic lengths of chromosomes of women is also more than that of men.

The chromosome 16, on which the beta globin gene is situated, is 202 centimorgans long in females and 132 centimorgans long in males, while the DNA molecule of chromosome 16 in men and women is of the same length ($\approx$100 million base pairs).

CHROMOSOME 16

Genetic linkage map of chromosome 16. Note the length of female versus male chromosome genetic maps.

In 1960, George Barski, Serge Sorieul and Francine Cornet, working in the Institute Gustave Roussy in Paris, developed a technique that gave a powerful tool to human geneticists. They fused somatic cells of two different strains of mice and successfully grew hybrid cells in culture. These hybrid cells had two important features. First, the two different sets of chromosomes contained by them remained functional and synthesized the proteins coded by genes of both the parents. Second, as the hybrid cells multiplied, some of the chromosomes were lost. Hence, hybrid cells with different combinations of the parental chromosomes (and genes) were produced. This procedure, 'somatic cell hybridization', was used to fuse human somatic cells.

Human cells, with the exception of germ cells, do not fuse easily. And it is still more difficult to make a human cell fuse with that of an animal. Specific chemicals, that act on the cell surface, break this barrier and are used to fuse human cells with mouse cells. However, the resultant hybrid cell is unable to carry the large number of chromosomes contributed by the two different cells. To survive, the hybrid cell resorts to shedding some load of its genome by dropping some chromosomes, quite randomly. Interestingly, most of the chromosomes lost are of human origin. This behaviour of cells has been ingeniously used by scientists to find out which chromosome carries a particular gene.

> Men and mice are a heady mix
> Hybrid cells are in a fix!
> Most fused cells do feel a bit funny,
> Chromosomes to be carried are sure too many.
> Shed some of them and take it cool,
> Of course, that gives us a mapping tool!

This is how it is worked out. A mouse cell is made to fuse with a human cell. The 'hybrid' cells, over a period of time, retain only a few human chromosomes. It is possible to get a cell with one or two chromosomes of human origin left in the fused cells. If these cells continue to synthesize a protein of human origin, it can be safely assumed that the gene for the protein is present on the chromosomes still retained by the hybrid cell. As there is no way of controlling the number or type of chromosomes lost in a somatic cell hybrid, a large variety of hybrids, each with different human chromosomes, are generated.

Hybrids producing a human protein are screened for the presence of a specific chromosome carried by them. If a particular chromosome is always present in all hybrid cells synthesizing the protein in question, the gene for that protein can be said to be present on that chromosome.

For instance, when human chromosome 17 was retained in a hybrid cell, it was seen to produce thymidine kinase. Conversely, no thymidine kinase was detected when the hybrids did not carry the chromosome 17. Logically then, the gene for the enzyme thymidine kinase was assigned to chromosome 17. Using this method of somatic cell hybridization, several hundred human genes for proteins, which could be detected with ease, were mapped to their respective chromosomes.

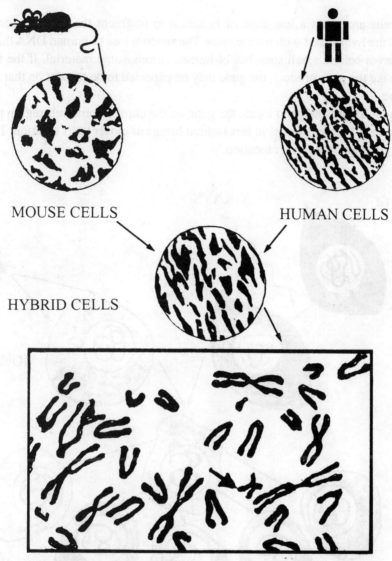

MOUSE CELLS   HUMAN CELLS

HYBRID CELLS

HUMAN CHROMOSOME (⟶) FROM HYBRID CELLS

Technique of somatic cell hybridization.

Over the years, the technique of somatic cell hybridization has undergone several refinements which have been extremely fruitful. The human chromosome is huge by molecular standards. To get closer to the gene, it was important to identify a much smaller stretch of DNA that could lodge the gene. A simple modification of somatic cell hybridization has been used to find out the sub-region of a chromosome where the gene is located. This technique is based on the effect of radiation on chromosomes. When chromosomes are exposed to radiation, they break up into small pieces.

Human cells are given a low dose of radiation to fragment their chromosomes. These irradiated cells are hybridized with mouse cells. The random loss of human DNA that ensues in hybrid cells leaves behind small stretches of human chromosome material. If the hybrid cell continues to make the gene product, the gene may be expected to be present on that small piece of chromosome.

Screening of hybrids helps to locate the gene on the chromosome sub-region present as a fragment in the cell. Mapping genes in this fashion brings us closer to its location. There is still a long way to go to get to its exact location.

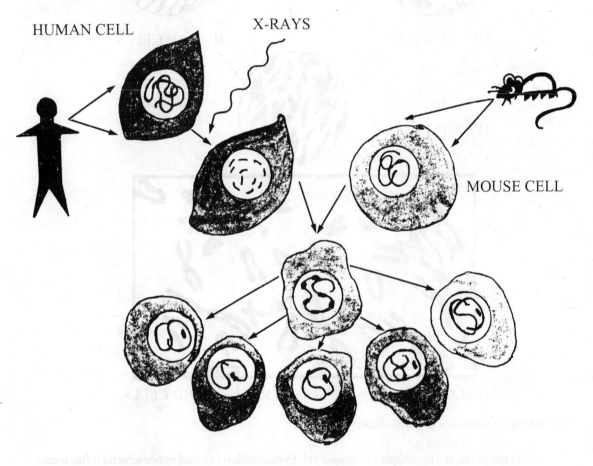

HUMAN CELL   X-RAYS

MOUSE CELL

SUB-CLONES OF HYBRID CELLS, EACH WITH A PIECE OF HUMAN CHROMOSOME.

Modified technique of somatic cell hybridization.

D. Bonstein from MIT, R. White and M. Skolonk of the University of Utah, and R. Davis from Stanford laid out rules of gene mapping based on simple logic. Let us use an analogy. How does one locate a house in a sprawling city? One of the ways is to locate it with respect to the 'landmarks' already known. Bonstein and others proposed a similar approach. They suggested identifying landmarks or 'markers' on the chromosomes to be used as reference points for locating a gene. Restriction sites were one set of these markers.

Little was known about such landmarks on human chromosomes. In 1960, it was observed that bacterial DNA has a number of sequences that appear to be repeated and are dispersed all over their genome. In the 1970s an explanation was given for these sequences, when a set of enzymes (Restriction endonucleases) was discovered, again in bacteria, which chewed up the DNA at some of these repetitive sites. These enzymes worked when such sequences were in foreign DNA, as in viruses. Such sites on their own genome were chemically protected. The microbes carried these enzymes as a part of their survival strategies.

List of some restriction enzymes which allowed scientists to cut DNA at well defined sites.

Restriction enzyme	Extracted from	Sequence recognised
Eco R1	*Escherichia coli*	G|A A T T C C T T A A|G
Hpa I	*Haemophilus parainfluenzae*	G T T|A A C C A A|T T G
Hae III	*Haemophilus aegypticus*	G G|C C C C|G G
Msp I	*Morazella species*	C|C G G G G C|C
Bam HI	*Bacillus amyloliquefaciens*	G|G A T C C C CT A G|G

Soon several types of restriction enzymes were discovered: those which chopped within the molecule, the 'endonucleases', in contrast to those that cut DNA from its two terminal ends, the 'exonucleases'. Some enzymes cut solely at exclusive sites, and were, therefore, called 'restriction endonucleases'. The points on the DNA molecule where the restriction endonucleases snipped, the **restriction sites,** became the landmarks that helped mapping the gene to a location on the chromosome.

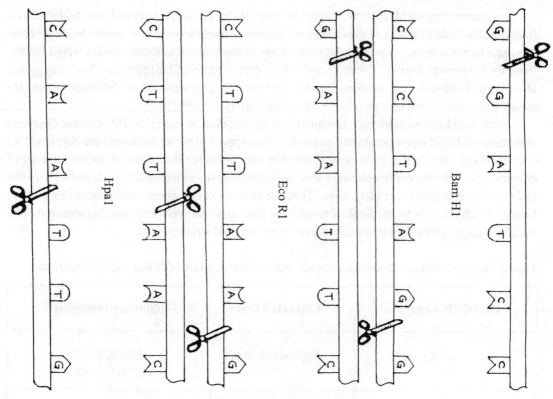

Snipping like a pair of scissors,
With a cut in the right place,
Fragmenting the DNA is the endonuclease.
The sizes of these bits
You just cannot miss,
If you follow the pattern of electrophoresis.
Treat it with a probe,
To highlight the band
And you have a mapping tool in hand !

Every time a restriction enzyme cuts the DNA, small molecular pieces of DNA called the 'restriction fragments' are produced. For instance, Eco R1 cuts the DNA molecule whenever it encounters the bases ACTTAAGA in the forward direction and snips between A and G. Hence, 'Eco R1 restriction fragments' are obtained after treatment with Eco R1. The resultant number of DNA fragments will depend on the number of restriction sites on the DNA. If the DNA has three such sites, it will break into four fragments.

The number of such cut fragments can be found out if the chopped DNA is made to travel through a medium (gel) in an electric field. This is the electrophoretic technique and separates out DNA fragments of varied lengths, depending on their sizes. The fragments are visualized as distinct bands on the gel. For instance, four bands seen on a gel after digestion with a restriction endonuclease implies that the DNA fragments under investigation have three restriction sites.

The number of landmarks on a chromosome which are obtained by use of one restriction endonuclease are limited. To create more landmarks, other endonucleases are used, and these are also of bacterial origins. Each one of them cuts the DNA at a specific restriction site. This way a host of landmarks are identified on the genome. Using several restriction endonucleases, the genome of a person can be fragmented and a variety of landmarks can be positioned. This is the restriction map of the genome of that person.

Just like a fingerprint, the restriction map of a person is unique. This is because each one of us has a specific number of restriction sites at variable distances in our DNA molecules. The variability in restriction sites shows up as variations in size and number of restriction fragments, and can be identified as a distinct number of bands in electrophoresis.

Fingerprints alone do not identify you
Markers are there in your DNA too.
To get your DNA fingerprints
A bit of your hair, or a drop of blood
Will surely do!

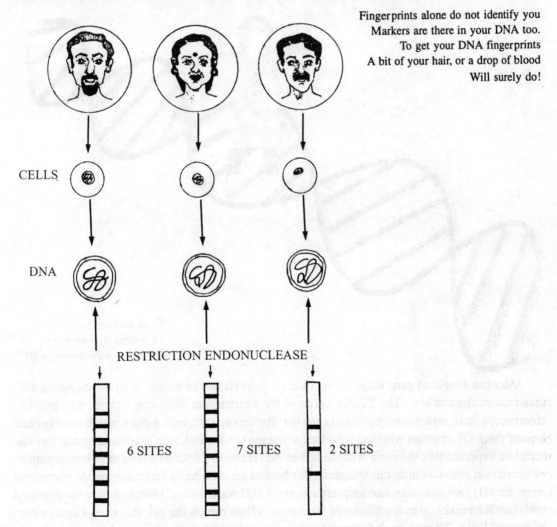

CELLS

DNA

RESTRICTION ENDONUCLEASE

6 SITES   7 SITES   2 SITES

Bands during electrophoresis

Understandably, the lengths of these restriction fragments are highly variable in our population. Hence, human population is 'polymorphic' with respect to DNA fragments generated by the use of restriction endonucleases. To put it simply, just as there is a large variety of facial features in a population, there is an equally large variety of restriction fragments which vary in their sizes and numbers in each individual. And just like the inheritance of our physical characteristics, such as the shape of the nose, or the 'diseased' gene from one of our parents, we also inherit these restriction sites: some from the maternal side and others from the paternal side. Hence, each one of us has a unique map of restriction fragments.

Gene mappers have found restriction fragment analysis extremely useful for tracking genes, as a new form of linkage mapping was possible!

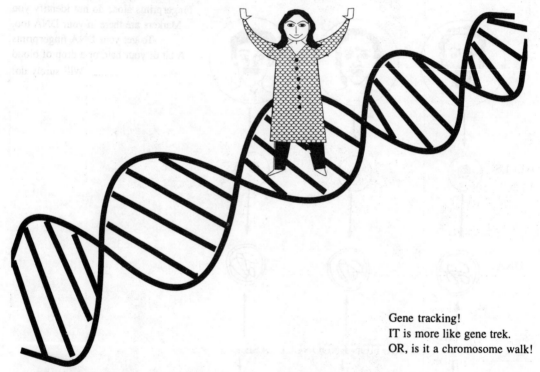

Gene tracking!
IT is more like gene trek.
OR, is it a chromosome walk!

Morgan's logic of gene mapping by linkage analysis can be extended to the mapping with these molecular markers. The DNA marker—here a restriction site—can be used as a point of reference, a landmark whose position is known. The linkage between a gene and the marker can be analysed. Of a person who has inherited a disease (gene) and, also inherits a restriction site it can be expected that the gene is close to that site. However, following the inheritance pattern of restriction sites is a little complicated. The human genome has a large number of restriction sites: Eco R1, for example, can snip after every 4,000 nucleotides. Hence, a genome digested with Eco R1 breaks up into millions of fragments. When run on the gel, this shows up as a blur of stained DNA. Though each fragment runs as an independent band, it can be seen distinctly only if it is highlighted. The highlighter is the 'probe' about which we talked earlier.

Gene for Congenital Cataract Identified in India

In India, a large number of children are born with congenital cataract in both eyes. This abnormality soon leads to impairment of vision. Prevention of congenital cataract and the ensuing loss of vision is possible only if the gene responsible for the disease is identified.

A major breakthrough in this direction has been made jointly by the scientists from Amritsar, Punjab, working with the Centre for Genetic Disorders of the Guru Nanak Dev University, and Dr Daljit Singh Eye Hospital, and those from the Institute of Human Genetics of Humboldt University, Berlin, Germany. They have succeeded in localizing the position of a new gene which is responsible for one type of congenital cataract (cerulean cataract with sutural opacities). This gene has been localized to the long arm of chromosome number 22 with the use of new molecular genetic techniques, involving a series of DNA markers called microsatellite markers.

The work, spread over many years, involved identification of a large number of families suffering from congenital cataract over several generations. The breakthrough was achieved through the cooperation of a five-generation family with more than 30 affected individuals. To identify new genes, it is imperative to have cooperative and well informed families, along with precise diagnosis of the disease.

In a city full of landmarks, the house to be located cannot be found unless the landmark nearest to the house is known. Similarly, to locate a gene, a specific landmark must be identified by highlighting it. This is what a probe does. Remember, a probe is a chain of nucleotides of known base sequences used to detect a complementary sequence in the DNA. To detect the presence of a probe, it must carry a label that distinguishes it from all other DNA molecules. Imagine a city with identical milestones all over its terrain. To identify the one of interest, a special sign can be put on the milestone to brighten it up. The probe is like a sign board which must be lit up to be noticed. A radioactive element or a fluorescent dye can light up a probe and, in turn, trace the bit of DNA it has hybridized with.

To begin with, the genome is isolated from a cell. The molecules of DNA are unwound and made single stranded by special treatments. The probe too is made into a single stranded molecule. When the two are mixed, the probe binds to a complementary region of the genome. As the base sequences of the probe are known, the corresponding sequences of the genome can be inferred. This technique has been modified extensively to study gene structure and function.

The hunt for the gene begins by trying to establish a connection between the piece of DNA lit by the probe and an inherited disease. In this effort, cooperation of members of the family where the disease is prevalent is of considerable help to scientists. Blood samples from several relatives are collected. To make meaningful analysis, often large families spanning at least three generations, are involved in this study. DNAs from these individuals are run on gel electrophoresis and the fragments of interest are visualized using the probe. Information about each of the

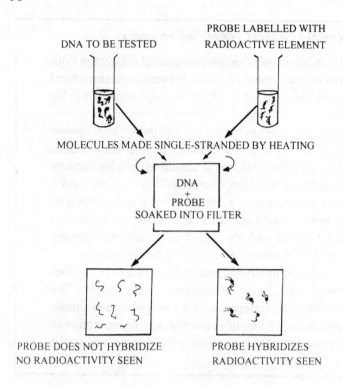

PROBE LABELLED WITH
DNA TO BE TESTED   RADIOACTIVE ELEMENT

MOLECULES MADE SINGLE-STRANDED BY HEATING

DNA
+
PROBE
SOAKED INTO FILTER

Labelled with a fluorescent dye,
Or with P thirty-two,
The probe is on a hunt and,
It has a job to do.
It has to track down the sequences
Wherever they may be...
The complementary bases,
To be lit for you to see.

PROBE DOES NOT HYBRIDIZE   PROBE HYBRIDIZES
NO RADIOACTIVITY SEEN   RADIOACTIVITY SEEN

Diagrammatic representation of a simplified procedure for using a
probe to detect the DNA fragment of interest.

family members regarding the disease and the presence of the DNA fragment, is put together. To ascertain the linkage between the disease and the DNA fragment, a computerized statistical analysis is carried out. Although a little complicated, this analysis is based on simple logic.

Two alternatives are considered. First, the disease gene and the DNA fragment are not linked, but they are assorted independently. Second option assumes that the disease trait is linked to the inheritance of the fragment being studied. The likelihood of obtaining the data as seen in the pedigrees, if each one of the two is true, is worked out. The ratio of the two possibilities provides a good measure to ascertain whether the disease is associated with the marked DNA fragment. Adding mathematical corrections related to numbers observed, geneticists calculate what is known as a lod score. A lod score of three is evidence for linkage of the disease to the highlighted DNA fragment.

A restriction fragment that has been highlighted by a probe is an identified physical position on the chromosome. If a gene that causes the disease is linked to a restriction fragment, it gives a very exciting piece of information. It identifies the physical location of the gene.

Travelling in an unknown city, with fragments of the map on a paper can be a nervous experience. But the actual landmarks of the map are more reassuring. Restriction fragment length polymorphism (RFLP) analysis is not just a way of positioning genes with respect to each other, but it is a powerful method to zero down to that bit of DNA that carries the gene.

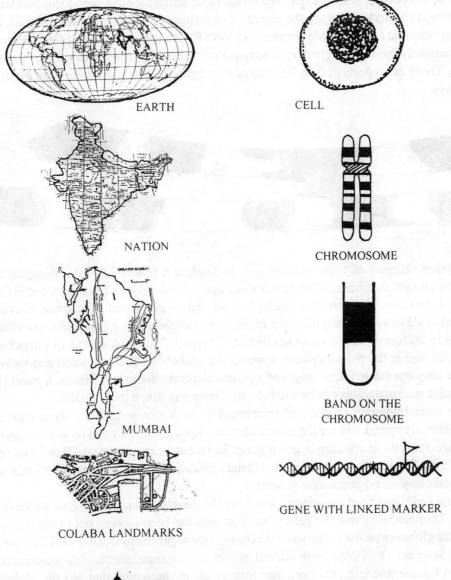

Locating a gene on a chromosome is analogous to mapping a site in a country.

In the last decade, several other sites on the DNA molecule have been identified which act as effective landmarks. These are the repetitive sequences of varying number of bases, and are known as 'variable number tandem repeats' or VNTRs. An extremely useful set of such repeats used extensively by human geneticists today are microsatellites (now also called short tandem repeats). These are repeats of 5 to 50 pairs of 3 to 6 nucleotides spread over the genome like milestones.

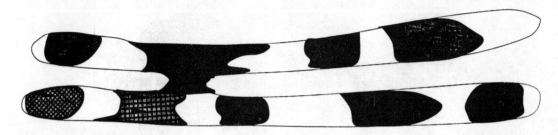

Another category of these markers are the tandem repeats of 10-60 nucleotide blocks. These are known as minisatellites. Both mini and microsatellites have replaced RFLPs as landmarks. Genome mappers now prefer to find their way around with these markers. As compared to RFLPs, microsatellites are highly polymorphic, i.e., a given microsatellite may exist in 4 to 20 forms, due to variable number of repeats. This high degree of polymorphism implies that higher the polymorphism, lower is the probability that two unrelated individuals will have the same allele (same length of a given microsatellite locus). Hence, a panel of such polymorphic microsatellites can be used to fingerprint any given individual.

The other important property of microsatellite is that over 6,000 of these markers are nearly uniformly spread all over the human chromosomes. Thus, they can be used as landmarks for linkage analysis in tracking a given gene. In recent years, microsatellites have become powerful tools for genetic analysis and have found applications in forensic investigation, paternity testing, and clinical and population genetics.

Vinita and Vinod need the information about the thalassaemia causing gene which they are carrying. They are lucky that the gene of their interest has been worked out extensively.

As the globin protein was known, it was easy to make probes to localize the DNA segments carrying the gene. The DNAs from normal as well as thalassaemia patients were studied for restriction fragment length polymorphism, especially in the region that has the globin gene cluster. This region could be studied as the probes for it were already available.

In a normal population, certain sizes of restriction fragments generated in the globin gene cluster were seen to be common. On the other hand, some of the restriction sites were exclusive to a person. This pattern of restriction sites in the globin gene region of normal individuals is technically called 'normal RFLP haplotype'.

Different sub-populations over the world have slightly differing restriction site patterns or RFLP haplotypes. But some restriction sites are shared with other populations. For instance, there are five restriction sites in the region of the beta globin gene that are common to the haplotypes of normal populations among Greeks, Italians, and Asian Indians. Once the normal haplotype is worked out, it is possible to find mutations that cause thalassaemia.

To understand the genetic differences within the Indian population and between peoples from other countries of the world, a programme was initiated in 1997 by the Department of Biotechnology (DBT), New Delhi. This programme will also help understand the relationship between genomic variation and the disease patterns in the Indian population. Some tribes of Andaman Islands, select groups from Kerala, and the Parsi community are of special interest in this study as they are susceptible to some inherited diseases. RFLP analysis and related techniques will be used in this study.

A change in the base sequence that disrupts a restriction site can be identified by RFLP analysis. Thus, several hundred mutations in a restriction site in the globin gene region have been identified. A base change in the globin gene may be such that it cannot be recognized by the restriction enzyme. A mutation could also add a restriction site. Change of a single base can be picked up by looking at the pattern of restriction fragments. The globin gene has a restriction site for one enzyme or the other, after almost every 500 base pairs. In other words, a mutation can be picked up at a distance of every 500 bases.

This approach can also be used to diagnose the type of defect causing the disease, with an appropriate probe. For example, in the globin gene region, adenine replaces guanine at the position 110. Now, two probes can be synthesized: one with base sequences complementary to

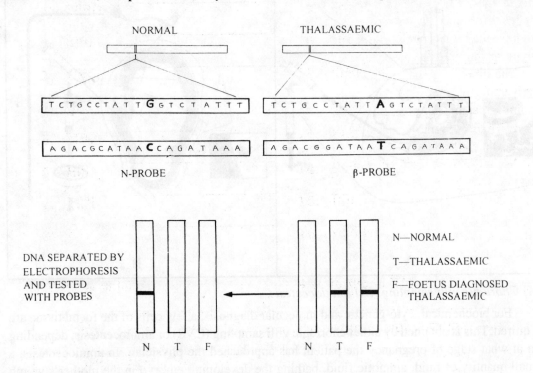

Use of probes for diagnosis of thalassaemia.

the normal sequence (G probe) and the other corresponding to the abnormal sequence (A probe). The DNA from the patient is collected, cut into restriction fragments and is then treated separately with the two probes. If the patient's DNA has the abnormal sequence, it will hybridize with the A probe, implying that he/she has an abnormal haplotype. This technique is now being used effectively and routinely for prenatal diagnosis of thalassaemia.

Vinita and Vinod can also avail of this technique to find out whether their second baby will be thalassaemic.

Detection of a defective gene in the foetus in early pregnancy gives the Vasvanis an opportunity to take a decision. Now they have a choice of terminating the pregnancy if their second baby has inherited the thalassaemic gene(s). And indeed, the baby's DNA is abnormal.

In general, prenatal diagnosis is done using several techniques. Ultrasonography or radiography techniques allow visualization of the foetus in the mother's womb with the least risk to the mother and the baby. The details of anatomy of the foetus, including gross abnormalities, are detected with ease. Similar information is also obtained by other techniques such as foetoscopy, foetography, and amniography. Genetic abnormalities leading to abnormal foetal anatomy are detected by these methods. In instances where the genetic defect is biochemical, at the DNA level, these tools are of little help.

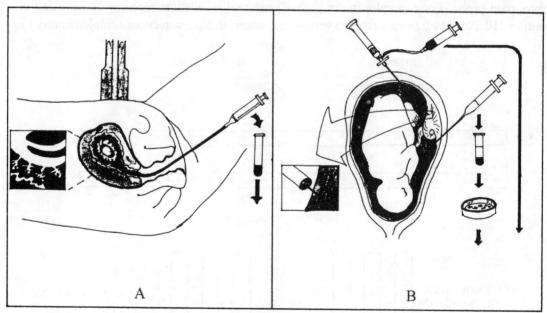

(A) Chorionic villi sampling; (B) Amniocentesis.

For biochemical, cyto-genetic and molecular diagnosis, a few cells of the foetal tissue are required. This is obtained by either chorionic villi sampling (CVS) or amniocentesis, depending on at what stage of pregnancy the patient has approached the physician. In amniocentesis, a small quantity of fluid, amniotic fluid, bathing the developing embryo in the mother's womb and containing a few foetal cells, is taken out in the fifteenth week of pregnancy. The cells in the fluid are first multiplied by culturing them and then tested. In chorionic villus tissue method, the cells of the membrane which intimately surrounds the embryo and contains only embryonic cells, are taken out and tested. This method is performed between the eighth and tenth week, and is considered safer than amniocentesis. DNA is isolated from such foetal samples. In this way, several genetically inherited diseases are identified prenatally using appropriate molecular probes.

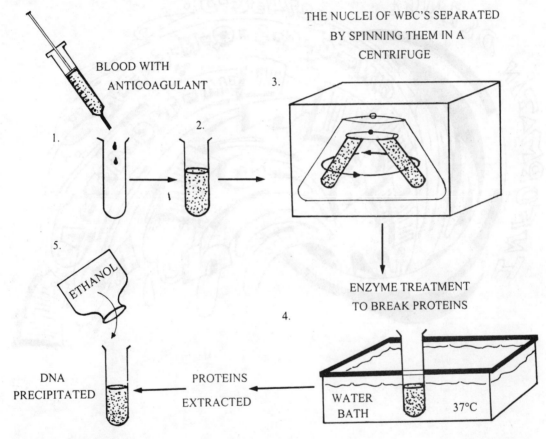

DNA isolation.

For Vinita and Vinod, it is now possible to know by prenatal diagnosis, especially by using the molecular approach, whether they have passed on the thalassaemic gene(s) to their second child. And the diseased gene was detected in the unborn baby at an early stage. One way of helping families with inherited diseases is by genetic counselling and advising early termination of pregnancy. Most mothers in several Mediterranean countries, and in the UK, avail of the prenatal diagnosis and termination of pregnancies of foetuses affected with thalassaemia and other similar blood disorders—technically known as globin or haemoglobin diseases. This is more so if they have already got one diseased child. These services are provided free by the State.

This benefit is also provided in India at several centres in the metros. A National Thalassaemia Control Programme is being launched in the country, jointly by the Indian Council of Medical Research and the Department of Biotechnology, New Delhi. This programme involves setting up of regional centres in different parts of the country, where DNA diagnostics and prenatal diagnosis, combined with medical care and genetic counselling are being carried out.

7 HUMAN GENOME PROJECT
Deciphering the genetic molecule

In 1543, Andreas Vesalius published, '*The seven books on the structure of the human body*'. Commonly known as the '*Fabrica*', this work was unique as it was based on actual observations of the body. Vesalian anatomy had a dramatic effect on contemporary medicine. A new way of looking at diseases emerged. Yet another revolution of similar nature is in the making. Another anatomy is being worked out which will change the way we look at ourselves and the diseases that afflict us. This is molecular anatomy.

All the twenty-three pairs of human chromosomes—the entire human genome—is being probed, all the genes are being mapped and each and every base sequence will be deciphered. The work is of monumental nature and requires a number of modern Vesaliuses to pool in their

talent and creativity. Several teams of scientists from around the world have joined in this effort of 'knowing' ourselves at the molecular level. This most exciting project of our times, often compared to the landing of man on the moon, is the 'Human Genome Project' (HGP). The outcome of the project is not only going to make a difference to Vinod and Vinita, but is also going to change the quality of our lives and the way in which we tackle diseases.

The 1970s and '80s have been an enriching period for molecular biologists. Several technical innovations took place that permitted the sequencing of small molecules of DNA of viruses. The possibility that the human genome too could be worked out in a similar manner was already in the air. But the enormity of the task was daunting, if not discouraging. Its scientific feasibility and its worth were debatable. Yet with time, it became clear that such a project was inevitable if we were to solve the major problems in biology, especially those regarding our health.

In 1986, The Department of Energy at Los Alamos, New Mexico, USA, organized a conference to discuss the possibility of such a project. In the following years, several committees were set up to assess the feasibility and usefulness of mapping and sequencing the human genome. As a fallout of the recommendations of these committees, the Office of Human Genome Research was created in September 1988 with J.D. Watson as its head. By January, the programme was laid out and the international initiative, 'Human Genome Organization' (HUGO) was formalized.

The inputs of the project were of a colossal nature. The estimated expenses were equivalent to that for a space mission. The technical inputs needed were huge. It was obvious that professionals from several disciplines—computer scientists, physicians, lawyers, social workers, and, of course, molecular biologists, all needed to pool in their expertise. Time was an important consideration of this entire project, if the efforts were to be worthwhile!

Time has always been an important factor in all our activities. More so, when collaborative, multi-disciplinary and interdependent activities are undertaken. The planners of the HUGO were extremely conscious of time-bound targets to be met by all involved. Five year goals precisely defining the objectives of the project were set.

To begin with, a complete genetic map with markers as milestones placed 2 to 5 centimorgans apart would be obtained. This would involve not only the actual linkage analysis in various families by known set of markers, but also finding new markers. Attempts were to be made to develop an infrastructure that would make mapping easier, reliable and, if possible, automated.

The second major task to be undertaken simultaneously was to physically find the locations of the genes. Merging the genetic and the physical reality of genes was not an easy proposition. It required generation of bits of DNA by cloning and screening the clones for their position on the chromosome. With most of the mapping done, the next phase would be to determine the base sequences of the chromosome. This indeed,is a herculean task! Methods to do so, available at the time the project was envisaged, were neither fast enough to meet the target nor were they cost-effective. Automatic and reliable sequencing technology was yet to be developed.

Simultaneously, geneticists started sequencing and mapping the genomes of other life-forms. These animals were chosen carefully. Majority of them were already under the scrutiny of molecular biologists. These organisms, bacteria, worms, yeast, fish, mice, were definitely more easy to work with. There was also a strong suspicion about similarities in the genome organization across species. And, indeed, comparison of the available molecular data of these animals with that of the human brings out a remarkable degree of overlap. Mapping of yeast or mouse genomes, for instance, and comparing them even with a very rough map of the corresponding part of the human genome, could provide starting points for locating human genes. And all these hopes have been more than fulfilled.

It was also anticipated that many more mapping and sequencing strategies would emerge to speed up human genome studies.

The Human Genome Project had another unique feature. Truly international in nature, sharing of information was obligatory among the participants and results had to be exchanged almost on a day-to-day basis. The vast data generated would be accessible to all involved. The development of a powerful information base and analytical methods that would interpret the data, was one of the most important goals of the project. Not only were computer-based systems to be developed, highly disciplined, hard working and motivated groups also had to be groomed to the massive task. Training and technology transfer was to be an integral part of the genome project.

Some model organisms (and their genome sizes) being worked on in the genome project:

Escherichia coli (4.6 Mb), a bacterium, the work-horse of geneticists.

Saccharomyces cerevisiae (14Mb), a budding yeast and an eukaryote, has been studied extensively by biologists.

Caenorhabdities elegans (100Mb), a round worm, its genome is of special interest to developmental biologists.

Drosophila melanogaster (165Mb), the fruit fly, is used for analysis of gene structure and function in developmental processes.

Fugu rubripes (400Mb), the puffer fish, is a model for studying vertebrate genome. Its genome has many exons which may overlap with those of the human.

Mus musculus (3,000Mb), the mouse is of interest to gene manipulators.

Rattus novegicus (3,000Mb), the rat is a model for complex vascular and neurological disorders.

Arabidopsis thaliana (100Mb), an angiosperm, is a representative of economically important plants which could be genetically manipulated.

It had not escaped the planners of the genome project that this new biology of human beings that they were working out, was going to bring about radical changes in the way we look at ourselves and our societies. Unusual pressures will have to be handled by individuals. What would it be like to know that you are a carrier of a killer disease ? How do you live with the dread of cancer when you carry the gene for its susceptibility? How will insurance companies and employers react to this information? Can we make designer babies free of all so-called 'bad genes'? Several other questions loom large, each a debatable one and to which answers are not simple. Generation of public awareness and addressing ethical and social outcomes is an important component of this massive project. Legal issues like patenting genes and gene products also need to be sorted out. These issues will have to be considered in the international context. The HUGO has plans here too.

Sooner or later, the genome project is likely to touch most aspects of our lives. But people like Vinita and Vinod will be the first ones to feel the immediate impact of HUGO. For identification of diseased genes, combined with their genetic and physical mapping, is high on the HUGO agenda. In fact, several experiments that primarily appear to be performed out of intellectual curiosity, have led to developing technical know-how for relieving human misery. Not surprisingly, the conceptual framework of this project rested on the outcomes of experiments carried out in the last two decades.

A technical problem that had to be solved before any biochemical analysis could be done was to get the gene (to be tracked) in sufficient quantities. A large number of cells would be needed to get enough of the gene. For instance, to get sufficient quantity of the beta globin gene, DNA from all cells of about two dozen persons would be required, and the gene would somehow have to be separated from the rest of the DNA. Gene isolation and manipulation seemed impossible until the 1970s. A technical break through ushered in a new era ! The development of molecular cloning methods and recombinant DNA technology made handling of genes easy.

In the early 1970s, Tom Maniatas and his colleagues from MIT constructed a 'genomic library'. How did they do this? To begin with, they took the complete genome of an individual and cut it into pieces using a restriction endonuclease. Each of these human DNA fragments was transported into a bacterium via viruses that infect bacteria (bacteriophages). As the bacteria multiplied, so did the piece of human DNA carried by it, producing unlimited copies of the human DNA fragments. A large number of fragments were created out of a single human genome. What had been created was a 'DNA or a genomic library'!

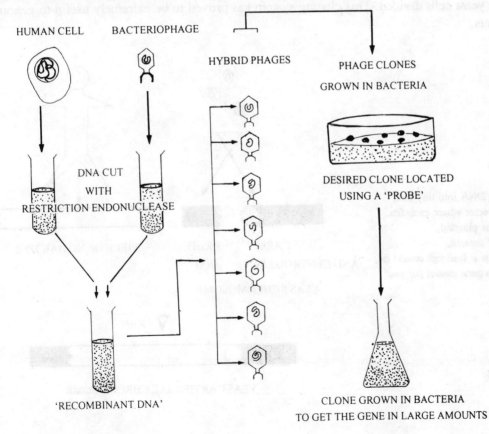

HUMAN CELL   BACTERIOPHAGE

HYBRID PHAGES

PHAGE CLONES
GROWN IN BACTERIA

DNA CUT
WITH
RESTRICTION ENDONUCLEASE

DESIRED CLONE LOCATED
USING A 'PROBE'

'RECOMBINANT DNA'

CLONE GROWN IN BACTERIA
TO GET THE GENE IN LARGE AMOUNTS

Making a DNA library.

The methods used in producing the DNA library were also used for amplifying (multiplying) bits of DNA. All that was needed to get more of the gene were its few copies. These copies were inserted into a 'cloning vector', like the bacteriophage.

There could be other vectors too. A 'plasmid', a little circle of extra-chromosomal DNA in bacteria, has the capacity to carry small DNA fragments. A designer vector, 'cosmid', has been made specially by combining a small piece of phage DNA with the plasmid DNA. A cosmid can carry large stretches of DNA. The gene was first tagged on to the vector DNA. This 'recombinant DNA molecule' was then inserted into a bacterial 'host cell'. As the bacteria grew, the gene fragments carried by them also multiplied. Millions of copies of any gene (β-globin, for instance), enough for analytical purposes, can thus be extracted from a bacterial clone.

For cloning large fragments of DNA, this technique was innovatively modified by making an 'artificial chromosome'! A recombinant molecule, **'the artificial yeast chromosome'** or **'YAC'** was constructed by combining the centromere and telomeres of the yeast chromosome with a human DNA fragment. These YACs were inserted into the host yeast cells and amplified as the yeast cells divided. This cloning system has proved to be extremely useful to genome mappers.

Cut the DNA into tiny bits,
Get a vector where each fits,
Phage or plasmid,
YAC or cosmid,
Now just a host cell would do,
To get a gene cloned for you!

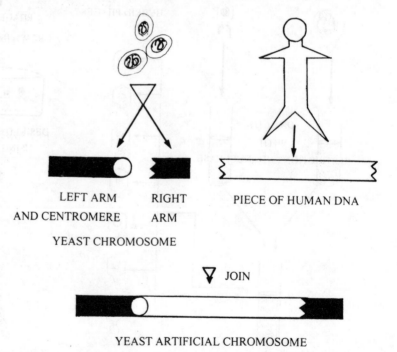

LEFT ARM   RIGHT   PIECE OF HUMAN DNA
AND CENTROMERE   ARM

YEAST CHROMOSOME

▽ JOIN

YEAST ARTIFICIAL CHROMOSOME

Artificial yeast chromosome.

In 1990, Kary Mullis reported another method of amplifying DNA. His idea was so remarkably simple that biologists now have often wondered why it was not thought of earlier! Mullis reconstructed all the conditions which are necessary for DNA replication within a cell in a test tube: the DNA to be replicated, the enzyme 'polymerase' that did the job, the 'primers' that initiated the synthesis of the new strand, and the DNA building blocks, the trinucleotides. A temperature controlled environment was provided to get duplication going.

The process started with the opening up of the DNA strands which were to be replicated. The primers attached themselves to the complementary site and helped the polymerase to get on with its job. The polymerase, using the trinucleotides, synthesized the other strand. Now there were double the number of strands to start the cycle of DNA synthesis. In no time, millions of DNA strands were made! An alternative to cloning, the polymerase chain reaction (PCR), is considered as a mini revolution in molecular biology. Now DNA can be handled in laboratories in an entirely different way!

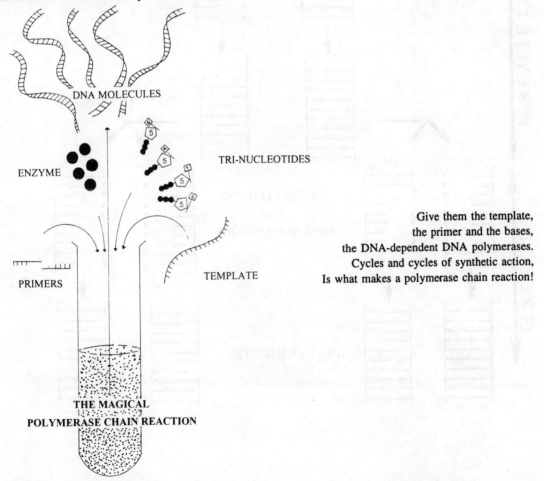

DNA MOLECULES

ENZYME

TRI-NUCLEOTIDES

PRIMERS

TEMPLATE

Give them the template,
the primer and the bases,
the DNA-dependent DNA polymerases.
Cycles and cycles of synthetic action,
Is what makes a polymerase chain reaction!

THE MAGICAL
POLYMERASE CHAIN REACTION

Polymerase chain reaction.

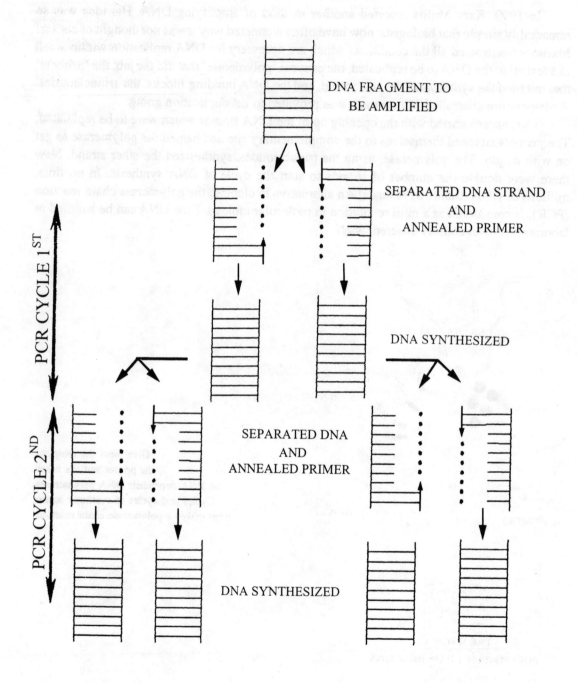

PCR amplification.

Equipped with the techniques of cloning and PCR, the human genome did not seem much of a puzzle. Indeed, both the techniques have been used ingeniously to map the genes and far more. The most rewarding outcome of cloning and PCR was the synthesis of a number of probes for locating specific base sequences.

The large human genome is unwieldy, quite difficult to analyse in its totality. Breaking it into segments of smaller size and obtaining these in large numbers makes the task much easier. These clones can be screened for the presence of restriction sites of interest or for identifying some DNA markers. Clones can also be used to find the sequence of the nucleotide bases of the DNA bit. In other words, several characteristic features of the DNA fragments can be identified, specially as a large number of copies of these fragments are available. In reality, these fragments are a part of the whole chromosome. Physically locating their sites onto the chromosomes will tell us the characteristics of that region of the chromosome. How can this be done?

The cloned DNA fragment to be located is labelled by a fluorescent dye or a radioactive molecule and used as a 'probe'. These labelled clones or probes can now be located on a specific chromosome by the technique of 'in situ hybridization'. The method is simple. A slide with a spread of chromosomes can be prepared using the lymphocytes from the blood. Each chromosome pair can be identified on the basis of its length and by the size of its two arms. The chromosomes at metaphase are tightly coiled. They can be treated to unwind them a little and then exposed to the probe.

The probe hybridizes in that region which has complementary sequences. The region where the probe has bound lights up as it carries a label. Fluorescence-labelled probes are popular for identifying the position of genes on the chromosomes. It is possible to get the entire chromosome lighted up by using multiple probes on the same spread of chromosomes. Called 'chromosome painting', this innovative technique is used for mapping several cloned fragments on to individual chromosomes. And if the cloned fragment happens to contain a gene of interest, its position on the chromosome can actually be visualized!

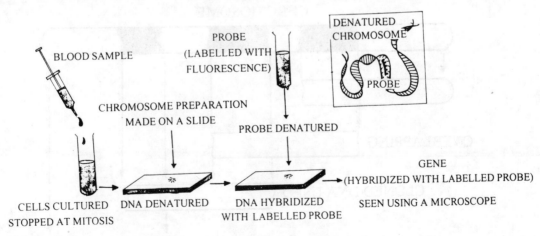

Technique of in situ hybridization.

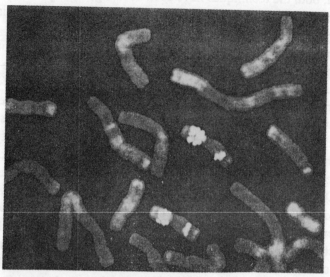

Metaphase chromosome spread.

The positioning of all the cloned fragments into their correct physical sequence-positions is what mapping is all about ! And that is what the genome mappers are out to achieve. They would like to reconstruct each of the chromosomes by positioning the fragments in place. This is like solving a linear jigsaw puzzle!

To be sure that the position of the fragments are correct, it is a good idea to work with overlapping fragments of DNA. This can be done if different endonucleases are used to cut the

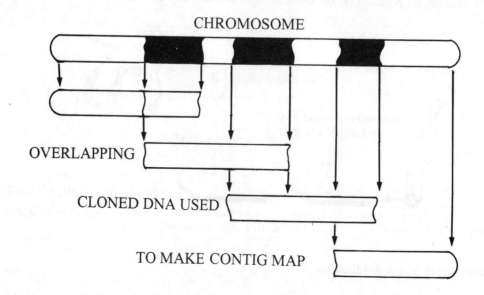

genome. A map of the chromosome with all its overlapping fragments in place is called its human 'contig' map. Once the cloned fragments fall in place, genome study becomes simple. All that is needed is to know each of the fragments well, right up to the sequence of its bases.

To make contig mapping easier and more precise, DNA clones from single chromosomes have been prepared. Elegant automated techniques have been developed to sort out each of the chromosomes. These DNA are then amplified using PCR. Further, they are cut and made into a set of clones. We can now get a DNA library of any chromosome. The library of chromosome 16 can be made and somewhere in this library is the globin gene, which is of interest to Vinod and Vinita. Once the globin gene is mapped and sequenced, there is a 'brave new' possibility for them!

The HGP, in combination with the increasingly powerful molecular biology tools, is expected to considerably clear up the cancer scene, too, especially cancers which run in families. The term 'cancer' stands for a group of diseases. A common characteristic of all cancers is the unruly manner in which the cells divide and move out from their positions and spread into other organs and tissues.

Two major sets of genes—**proto-oncogene** and **tumour suppressor (TS) genes**—are involved in triggering cancer. The first set, proto-oncogenes, codes for proteins which are part of a cell's signalling repertoire. Hence, a mutated proto-oncogene, or an oncogene, disturbs the signalling mechanism and stimulates uncontrolled growth. The other set of genes, tumour suppressor genes, codes for proteins which restrain cell division and tissue growth. p53 gene (17p13) coding for the p53 protein, is one of the major TS genes, restraining cell division and inducing damaged cells to commit suicide. It is a versatile gene involved in cell cycle control, repair of DNA damage, initiation of programmed cell death and ageing. Inactivation or malfunction of this gene can lead to instability of the genome, loss of cell cycle regulation leading to cancer. There is strong evidence of the involvement of p53 in carcinogenesis but details are yet to be worked out.

Our cells have also evolved several types of back-up gene systems to take care of some stray cells which might escape the numerous regulatory controls. Mutations in these back-up genes stabilizes the roving cancer cells. Normal cells have several other mechanisms for repairing damaged DNA. Some mutations affect these DNA repair mechanisms, with the resulting mismatch repair predisposing cells to cancer. Melanomas or skin tumours, for instance, are caused by mutations induced by UV light, with the proof-reading enzymes failing to correct these errors. Hereditary non-polyposis colon cancer and neurological degeneration and a variety of tumours are two other instances of diseases caused by inactive proteins affecting DNA repair mechanisms.

The genetics of cancer is complex and often more than one genetic change is responsible for the onset of the disease. Majority of these changes—mutations—take place in somatic, mitotically dividing cells, due to exposure to environmental agents, including viruses. Certain food items and habits increase cancer susceptibility, with the disease showing up in middle age or late in life, when accumulated mutations trigger the progression of cancer. In some, however, the disease onset is early. Often these individuals have inherited mutations. In such persons, cancer is said to be inherited and the mutations predispose an individual to a cancer of a particular type. Scientists have identified several inherited mutations associated with cancers.

(1) Colon cancer is caused by a mutation of TS gene
(APC-5q21), leading to colonic polyps. Individuals with
this mutation are predisposed to colon/stomach cancer.

> *(2) Mutations in two TS genes, BRCA-1 (17q21) and BRCA-2*
> *(13q12-q13), are associated with high risk of breast cancer.*

(3) In men, mutations in BRCA-1 and BRCA-2 increase
susceptibility to prostate cancer.

> *(4) Kidney cancer is caused by inherited mutations*
> *of TS genes, WT-1 (11p13) or VHL (3p25-p26).*

(5) Mutated TS genes, MTS 1 (9p21) or CDK-4,
predispose individuals to skin cancers.

> *(6) Neuroendocrine cancers are associated with inherited mutations of TS genes NF-1*
> *(17q12-q22) or NF-2 (22q12).*

The completion of the HGP will unravel the structure of more mutated genes which cause different cancers. Molecular understanding of familial cancers would help high-risk individuals to take preventive measures by modulating diet and habits.

8 GENE-BASED THERAPIES
New trends in treatment of inherited diseases

There is great hope among scientists and clinicians that information about our genes—their functions, structure, and exact location on the chromosomes—will eventually lead to the treatment of several inherited diseases, including thalassaemia. This was one of the rationales for undertaking the daunting task of mapping and sequencing the genes. Patients and clinicians alike, are looking for new methods of healing. Indeed, the possibility of manipulating genes for therapeutic purposes promises to revolutionize the medical sciences. Obvious and immediate candidates for gene therapy are inherited diseases apparently caused by single genes, such as cystic fibrosis. Though caused by a single gene, these diseases have a widespread effect on the general physiology of an individual and often, more than one organ system is affected. The technology developed to date fails to handle such complications.

While the potential is unlimited, it is not easy to work out therapies involving our genes. As we get closer to the bare details of the hereditary molecule, its complexity hits us hard. After all, gene manipulations may not be that simple. And it will have to be done with caution. Hence, several strategies for gene therapy are being tried out, many on animal models and some on humans.

Secondly, the manner in which genes function has turned out to be highly complex. Genes cooperate with each other, control each other and often function as a team. Genes interact with

List of genetic diseases that are candidates for gene therapy.

Defective gene	Disease
Adenosine deaminase	Severe combined immunodeficiency
α1-Antitrypsin	Emphysema
Arginosuccinate synthetase	Citrullinemia
CD-18	Leukocyte adhesion deficiency
Cystic fibrosis transmembrane regulator	Cystic fibrosis
Factor IX	Haemophilia B
Factor VIII	Haemophilia A
α-1-Fucosidase	Fucosidosis
Glucocerebrosidase	Gaucher's disease
β-Glucoronidase	Mucopolysaccharidosis type VII
β-Globin	Thalassaemia
β-Globin	Sickle-cell anaemia
α-1-Iduronidase	Mucopolysaccarridosis type I
Low density lipoprotein receptor	Familial hypercholesterolemma
Ornithine transcarbamylase	Hyperammonemia
Purine nucleoside phosphorylase	Severe combined immunodeficiency

proteins too. These subtleties of gene expression have to be understood before therapeutic strategies can be planned. For instance, the synthesis of beta globin gene is tuned to the synthesis of the alpha globin gene, both situated on different chromosomes, such that the assembly of the complete haemoglobin molecule is optimal. This fine tuning of the two genes has to be achieved when gene therapy is attempted for beta thalassaemia. Much more about what regulates beta globin gene needs to be known before gene therapy can be attempted safely in human beings. Gene therapy may not work for inaccessible systems. For instance, the neural system is not easy to reach. Such systems may have to be treated during foetal development by gene therapy. Despite these problems, advances are being made in gene therapy.

Four year old Ashanti DeSilva participated in the first ever gene therapy. Ashanti suffered from 'Severe Combined Immunodeficiency Syndrome' (SCID). The immune cells defending her body from infection could not function normally because they could not synthesize the enzyme, adenosine diaminase. The painful childhood she expected was converted into an almost normal one when W. French Anderson, the Director of Gene Therapy Laboratories at the University of South California School of Medicine, and his team successfully tried gene therapy on Ashanti.

The protocol followed by the doctors was simple. They took some of the abnormal immune cells from her body and introduced the gene for synthesizing the enzyme adenosine diaminase into them. The altered cells were then injected back into Ashanti's blood stream. The strategy worked wonders and Ashanti had her genetically corrected immune cells taking care of her infections. This approach where a cell is provided with a normal gene to compensate for the genetic defect is technically called augmentation therapy. A number of single gene diseases can be overcome by augmentation therapy. However, inserting a gene into the cell is not an easy job. The gene must be in the right place in the host chromosome and it should work well to produce the protein in the required quantities.

Genes can be made to enter the cells by a number of methods. It is possible to simply push the specific piece of DNA through the cell membrane using a fine syringe. The cell membranes can be made more permeable to DNA molecules by treating them with special chemicals or by giving them mild shocks. Once inside, the bit of DNA is expected to find a place for itself on the genome. Often this does not happen. Genes get inserted in the right place and start functioning in only one in a thousand cells. A far more successful procedure is to insert genes in place by using carriers or 'vectors'. Viruses are extremely efficient vectors.

Viruses have a set of genes enclosed in a coat of proteins. Left to themselves, viruses are quite helpless. But inside another cell, the host cell, they can become active. They use their own genetic material and that of the host cell to replicate. Some viruses can even insert their genes into the genome of the host cell and continue to stay on in the host cell as 'proviruses'.

A type of virus, retrovirus, is a good vehicle for carrying extra genes and placing them in the right position into the host genome. For instance, designer retroviruses carry therapeutic genes instead of their own genes. But the possibility of an insertion of the gene in a wrong place and the resulting unregulated protein synthesis has restricted their use in gene therapy. However, gene therapy using retroviruses is performed 'ex vivo', that is, they are used to transfer therapeutic genes into cells removed from the patients' body which are then put back, just as in Ashanti's case.

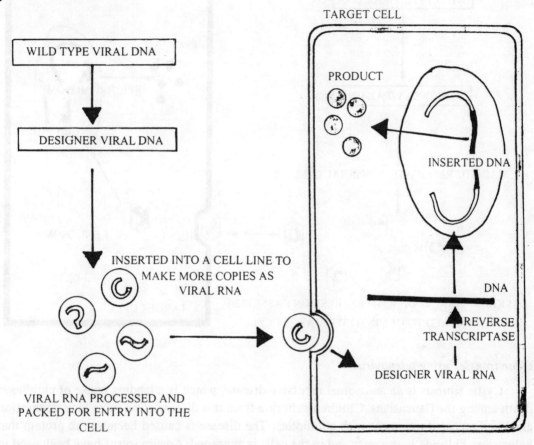

Diagram showing use of retroviruses for gene therapy.

Adenoviruses are another set of gene vectors. They carry the genes into the host cells but these genes do not get integrated with the host genome. Instead, the inserted piece of DNA stays afloat in the host cytoplasm as an independent identity. There is little risk of incorrect gene insertions into the host genome. A working gene set—a gene cassette—with the therapeutic genes and other stretches of DNA that regulate it, remains as an extra chromosome, the epichromosome. This epichromosome produces the missing protein.

This vector works well for 'in vivo' transfer of genes. The designed adenoviruses with the epichromosome, injected into the body, are expected to enter the host cells and synthesize the missing protein. As the epichromosomes are loosely associated with the genome, they are often lost by the host cell. Repeated therapy needs to be done to maintain the level of the therapeutic protein. Often, the host body reacts to repeated viral exposures and stimulates antiviral machinery, limiting the success of the therapy!

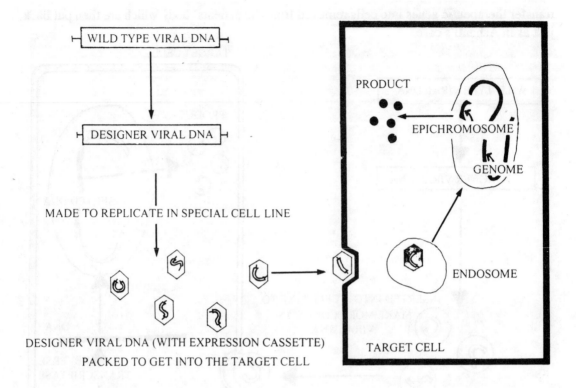

WILD TYPE VIRAL DNA

DESIGNER VIRAL DNA

MADE TO REPLICATE IN SPECIAL CELL LINE

DESIGNER VIRAL DNA (WITH EXPRESSION CASSETTE)
PACKED TO GET INTO THE TARGET CELL

PRODUCT

EPICHROMOSOME

GENOME

ENDOSOME

TARGET CELL

Gene transfer using adenoviruses.

Cystic fibrosis is an autosomal recessive disease, which is a leading cause of childhood death among the Caucasians. Children suffering from this disease often have lung congestion and need repeated treatment with antibiotics. The disease is caused because the protein that helps carry chloride ions across and in the cells is abnormal. Adenoviruses have been used to carry therapeutic genes to the lungs in these patients. While there is no promise for a permanent cure, this approach is of certain help to the diseased.

Several alternative procedures of gene transfer not involving viruses are also being tried out. Synthetic vesicles made of lipids, 'liposomes', are used as carriers of DNA. The liposome-mediated transfer of genes work well, as liposomes are non-toxic. However, for long-term relief, repeated treatments are necessary. The success of gene therapy depends a lot on innovative gene transfer protocols!

Scientists have recently developed a mammalian artificial chromosome (MAC) which will help to overcome problems related to gene delivery systems. These artificially designed chromosomes are expected to behave like natural ones in our cells. MAC-based vectors for gene therapy may soon replace YACs and other vectors, heralding an exciting phase for gene-based therapies.

MAC BRIGHTENS UP THE GENE THERAPY SCENE

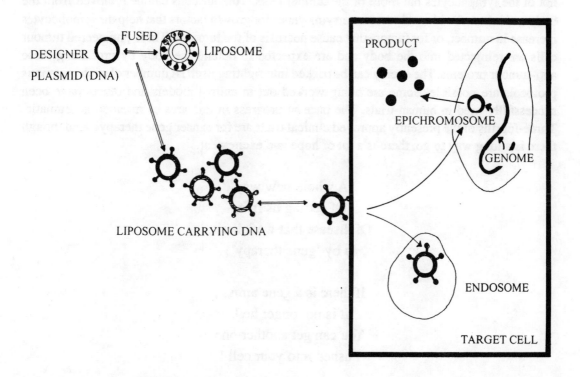

Gene transfer using liposomes.

Gene-based therapy is also being attempted for diseases with complex genetic bases, where more than one gene is implicated in the appearance and progression of a sickness. Cardiovascular diseases, hypertension, cancer, and arthritis are some such ailments. Therapies based on molecular nature of the gene are being attempted in such complex diseases.

The human body puts up a fight against any invaders. The body also resists any unwanted growth. Tumours are attacked by cells that defend the body, the lymphocytes. These lymphocytes put up a fight as the tumour grows. If the lymphocytes are successful, the tumour regresses. Often the tumour overwhelms the lymphocytes. Why does this happen? Why do lymphocytes fail? Understanding the basic biology of genes may help answer some of these critical aspects of tumour growth, and provide gene-based approaches to cancer management. Experiments involving gene transfer protocols have helped to understand the role of lymphocytes, especially those which infiltrate the tumours. To identify these tumour infiltrating lymphocytes, special marker genes can be inserted into them. This will allow us to follow the fate of these cells and understand how they build up into a cell population committed to tackle the tumour mass. This understanding will allow genes to be manipulated for tumour destruction.

Some innovative protocols involving gene therapy can be planned if the biology of cancer is well understood. One such protocol is immunotherapy which involves engineering genes, not of the lymphocytes but those of the tumour cells. Tumour cells can be removed from the body and infected with viral vectors carrying genes for growth factors that help the lymphocytes increase in number, or for factors that cause necrosis of the tumours. These engineered tumour cells are reinjected into the body and are expected to defend the body by producing these anti-tumour proteins. The enemy can be tricked into fighting itself! A number of such ingenious protocols are possible. Some are being worked out in animal models and others have been successfully tried in human trials. The pace of progress in this area of research is dramatic. Three-fourths of the presently approved clinical trials are for cancer gene therapy! And though there is a long way to go, there is a lot of hope and excitement.

A whole new way,
Of setting us free,
Of disease that we inherit,
Is by 'gene therapy'!

If there is a gene amiss ,
It is no longer hell,
You can get another one
Pushed into your cell !

Want your baby bright?
Sure, you can have a say!
From designer jeans
To designer genes
We have come a long way!

But should we play God?
And dabble in our well evolved design?
Are our fears misplaced?
Or will we have our Frankensteins?

Cardiovascular diseases are the major killers in urban areas. A majority of these problems are due to clogging of blood vessels. At present, blood vessels are grafted and reorganized using surgical methods. But formation of blood clots adds to the complications. Anti-coagulants like 'tissue plasminogen activator' can prevent the clotting of blood.

Such anti-coagulation factors can also be delivered to the required site by engineering the cells of the grafted blood vessels. These techniques are under investigation.

Acquired immunodeficiency syndrome (AIDS) is also a potential candidate for gene therapy. The disease is caused by retroviruses, the Human Immuno Deficiency Viruses (HIV), which lodge in the cells—lymphocytes—that defend the body against infections. These lymphocytes eventually die. For the virus to get into a cell, a molecular password is necessary. The virus must bind to the 'CD4 molecule' on the cell surface to gain entry. In other words, all lymphocytes bearing the CD4 molecules on their cell membrane are easy targets for HIV infection.

Now these lymphocytes can be removed from the body and genetically engineered to make them competent to fight the HIV onslaught! One of the possibilities is to change the lymphocytes genetically, such that they produce the CD4 protein and throw it outside the cells rather than carrying it as a membrane molecule. If a lot of the CD4 protein is in circulation, it would mop up the virus while the lymphocytes would be spared. Another possibility is to manipulate lymphocytes to produce proteins that block viral replication. While several such strategies are being tried out specifically to fight HIV, quite a few are also likely to be effective for other viral infections.

Imaginative gene therapy and gene-based therapeutic procedures are changing the clinical scene. Some diseases that are genetically better understood and less complex have been the choice for manipulation, others await breakthroughs!

For Vinod and Vinita, it might be a long wait. The haemoglobin gene is not easy to regulate and gene therapy for thalassaemia has met with little success. Yet, on the other hand, the quality of prenatal screening has improved tremendously and a number of thalassaemic mutations can be identified.

Several centres in the country provide the facility of identifying thalassaemic mutations based on molecular techniques. One of the techniques used is Amplification Refractory Mutation System (ARMS), which relies on polymerase chain reaction. For this reaction to take place, the strand of DNA to be amplified must complement exactly with the 'primers'(bits of DNA to initiate DNA amplification). If the 'primer' is mismatched because of a mutation, no polymerase

NORMAL   SINGLE BASE MUTATION

5'------————————— GGA ---- 3' DNA   5'------————————— TGA ---- 3' DNA
              | | | | |
        \ CT ---PRIMER   ◄——— ACT ----PRIMER
       A

UNMATCHED PRIMER   MATCHED PRIMER

Amplification Refractory Mutation System.

chain reaction would take place. On the other hand, if a 'primer' complementing a mutated (thalassaemic) site is provided to this DNA, it will replicate, and can be detected.

To check for a mutation, the DNA of the patient (or foetal tissue) is extracted and provided with a variety of primers, each of which matches different mutant DNA fragments. If the mutation is present in the patient sample, the specific primer will complement the DNA and initiate the polymerase chain reaction. The amplified DNA can be detected by electrophoresis. Several laboratories in India have developed the facility to identify dozens of thalassaemic mutations.

WHAT ALL CAN THE 'GENE GENIE' ACHIEVE?

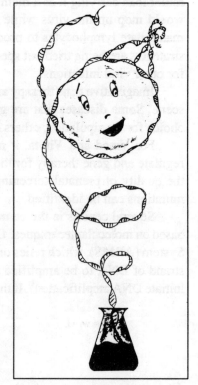

1) Control metabolic disorders...
by genetic manipulation of cells
    for synthesis of missing protein.
2) Rectify monogenic disorders...
by introduction of 'corrective'
    genes in somatic or germ cells.
3) Control vascular diseases...
by manipulation of cells to
    synthesize therapeutic agents.
4) Manage degenerative neurological disorders...
by grafting engineered cells that
    restore the neural functions.
5) Support cancer therapy...
by boosting the immune response.
by evoking cell-suicide.
by modifying tumour suppressor genes.
6) Increase the capacity to fight
infectious diseases...
by genetically modifying immune cells.

Gene manipulation has been tried out on somatic cells with an objective of relieving patients of their discomfort and pain. But no germ line therapy or manipulation of early zygotic cells has been attempted. While this may be the answer to some diseases, the risks involved are too serious. Moreover, there are moratoria on use of human embryos for experimental purposes in several countries.

Germ cells of other animals have been manipulated and it is possible to do so in humans. This raises a number of ethical and uncomfortable questions. It is imperative that we prepare ourselves to face the social change that genetic technology is likely to bring about. The biology involved is simple, the issues raised are shattering!

9 HUMAN GENETICS—PROMISES, HOPES, WORRIES
Education for preventing inherited diseases

The young Vasvani couple is overwhelmed with the advances made in human genetics. Their newly acquired knowledge has helped them to make up their minds. They are quite reconciled about the decision to abort the next pregnancy as prenatal diagnosis has revealed the second baby to be a thalassaemic. It was not easy to arrive at this decision. But the supportive and professional counselling had helped a lot.

Counselling, prenatal diagnosis and genetic screening hold out great promise, not only for an individual, but also for a society. However, these issues involve several moral and ethical concerns, which have become debatable and need to be carefully discussed by the scientific community and policy makers.

The job of genetic counsellors is to convey the medical and genetic facts in a simple manner so that it is understood by those who seek their guidance. While discussing the problems over a period of time, counsellors should not leave the affected mid-way. Support should be given both at intellectual and emotional levels, till a decision, one way or the other, is taken by a client. Persons seeking counsel could easily become shattered with adverse news.

The counsellors, besides being experts in both medicine and genetics, should be patient, sensitive and respectful towards their clients. On the other hand, counsellors could be driven to a wall if the affected persons become withdrawn or hostile. Indeed, there are several instances of in-laws pressurizing young couples to discontinue their visits to the counsellor!

Inherited diseases in families give rise to untold grief and misery, often resulting in divorce among couples, or depression, despair, and a feeling of guilt among the parents and the affected. These volatile moods need to be considered by counsellors whose attitudes must be warm and understanding.

Counselling is a tough job. Hence, counsellors are known to work as a team consisting of a clinician (with a strong professional knowledge of genetics), a geneticist (with a mathematical and statistical background), laboratory personnel to carry out varied tests, and a genetic associate or a trained medical social worker who could work easily among family members.

In India, counselling centres are attached to several public hospitals in major cities. These centres have facilities for diagnosis and counselling for inherited diseases, such as thalassaemia and other blood disorders, fragile X, DMD, albinism, cystic fibrosis, neuro-degenerative disorders, and mental retardation.

Genetic counselling is specially recommended for those couples whose one child is already affected with an inherited disease, as is the case of the Vasvanis. The presence of an inherited disease among the siblings and/or repeated miscarriages in a woman should prompt individuals to go for counselling, preferably before marriage and positively when a woman becomes pregnant. Couples whose ages exceed 30 years and are planning to have a baby, are greatly benefitted by counselling. Those with no family history of inherited disorders may also stand to gain, more so if the husband and wife are related to each other. Counselling is also urged for couples belonging to certain ethnic groups or castes.

Making a correct medical diagnosis is the most important step. A wrong diagnosis could be devastating for individuals concerned. The diagnosis involves a study of the chromosomes (karyotyping) to begin with, followed by biochemical studies and DNA analysis, wherein restriction enzymes and DNA probes are used. Diagnosis becomes complex for certain diseases caused by multiple genes.

A counsellor becomes doubly sure of the diagnosis if a disease is observed in three to four generations of a family. Care has to be taken not to miss out miscarriages, still births or relatives with mildly affected symptoms.

What are the chances of my baby carrying the disease-gene?

The answer is simple: 25 per cent chances for autosomal recessive genes where both parents carry a single copy of the disease gene; and, 50 per cent chances for autosomal dominant genes.

Today it is possible to eliminate the above statistical uncertainty, thanks to the various molecular tests. These tests precisely indicate the risk to the foetus which could be either 0% or 100%. Tests for several inherited diseases are available and their numbers are increasing rapidly as newer genes are being discovered! One predicts the risk of getting a disease by checking out the closeness of the marker RFLP to the disease-gene. A series of disease genes can also be directly checked out using DNA probes which are complementary to the different alleles of the gene.

At all stages of counselling, the risk of a disease showing up with every fresh pregnancy has to be emphasized.

Inheritance of a genetic condition is a random event, depending on which of the chromosomes of a pair from a parent is passed to an egg/sperm to form the embryo. Recall meiosis and Mendel's laws.

Some diseases where the genes are detected with DNA probes or mapped to nearby RFLPs and other DNA markers. (Ref. 1)

Disease	Chromosomal Location	Direct Detection or Approximate Map Units from DNA Marker
Adrenal hyperplasia	6 short arm	Direct
Agammaglobulinemia	X long arm	Marker, very close
Alzheimer's disease, familial	21 long arm	Marker, very close
Chronic granulomatous disease	X short arm	Marker, very close
Colour blindness	X long arm	Direct
Cystic fibrosis	7 long arm	Direct
Fragile X mental retardation	X long arm	Direct
Haemochromatosis	6 short arm	Direct
Haemophilia A	X long arm	Direct
Haemophilia B	X long arm	Direct
Huntington's disease	4 short arm	Direct
Hypercholesterolemia, familial	19 short arm	Direct
Lesch-Nyhan syndrome	X long arm	Direct
Muscular dystrophy, Duchenne/Becker	X short arm	Direct
Myotonic dystrophy	19 long arm	Direct
Neurofibromatosis	17 long arm	Direct
Phenylketonuria	12 long arm	Direct
Retinitis pigmentosa	X short arm	Marker, 8 map units
Retinoblastoma	13 long arm	Direct
Sickle-cell anaemia	11 short arm	Direct
Thalassaemia, α	16 short arm	Direct
Thalassaemia, β	11 short arm	Direct
Wilms disease	11 short arm	Direct
Wilson disease	13 long arm	Marker, 3-7 map units

Thalassaemic children celebrating World Thalassaemia Day (1997) with the noted ghazal singer, Pankaj Udhas.

Genetic counselling poses serious problems in India due to widespread illiteracy and general ignorance about genetic diseases. For instance, at one of the counselling centres in Mumbai, out of the recorded 600 to 700 families with thalassaemia major children, not a single family was aware about the disease before the birth of the child!

Field work is another important aspect of counselling. Social workers, assisted by medical practitioners and respected members of the community, can help create awareness of the disease in areas where the high-risk segments of the population live. Camps can be organized in schools and colleges. For instance, the Research Society of Bai Jerbai Wadia Hospital for Children and the Institute of Child Health, Mumbai, has established two satellite centres: at Ulhasnagar, a suburb of Mumbai with a population of 6 lakhs and 65 per cent of which are Sindhis, and at Rajkot in Saurashtra, where several high-risk Gujarati communities live. Screening laboratories have been established at these places, along with training of scientists and social workers to conduct screening tests and genetic counselling.

The following groups are offered counselling for thalassaemia:
(i) Individuals belonging to high risk communities like Sindhi, Lohana, Khoja, Jain, Schedule caste, etc.
(ii) Relatives of a thalassaemia major child.
(iii) Carriers, or thalassaemia minors or traits with or without a family history.
(iv) Couples at risk.
(v) Married and unmarried individuals with trait status.

Counselling is offered in different languages with counselling cards. Educational, economic, social, and cultural backgrounds, combined with traditional customs, psychological aspects, and social attitudes of the family, are all taken into consideration.

Several medical and social bodies are involved in creating awareness about thalassaemia in India. Some of these are affiliated to the Thalassaemia International Federation.

Some Thalassaemia Associations in India

Thalassaemia & Sickle Cell Society of Bombay, Vijay Sadan Flat 1, 168B Dr Ambedkar Road, Dadar TT, Bombay 400 014. *Tel. 22-414-2272/414-4453* *Fax. 22-414-0058*	Thalassaemic Children Welfare Assn. (Regd), 3047, Sector 20-D, Chandigarh. *Tel. 541032-39*
Thalassaemia & Sickle Cell Society of Bangalore, Jai Rattan Nivas, 11 A/22 Cunningham Road, Bangalore 560 002. *Tel. 258 661*	The Thalassaemia Society of India, 1/B Grant Street (near Chandni Market), Calcutta 700 013. *Tel. 2440656*
Parents Association, Thalassaemic Unit Trust, Kanji Khetshi, New Building, 3rd floor, T.G. Marg, Bombay 400 004. *Tel. 386 3347*	Research Society, B.J. Wadia Hospital for Children, Acharya Donde Marg, Parel, Bombay 400 012. *Tel. 22-413-7000/412-9786-7*
Thalassaemics India (Regd), C-1/59 Safdarjung Development Area, New Delhi 110 016. *Tel. 11-661199/6845461* *Fax. 11-6855721/6462970*	National Thalassaemia Welfare Society, KG 1/97 Vikas Puri, New Delhi 110 018. *Tel. 11-5507483* *Fax. 11-5591202/5598879*

Counsellor, Mrs H. Yagnik with her patients.

Detailed pedigrees as prepared by the B. J. Wadia Hospital for Children, Mumbai. These pedigrees reveal the number of people afflicted with thalassaemia, before and after genetic counselling (BGC and AGC).

Case No. 1 (File no. 115): Two carriers gave birth to two children. BGC: One child is thalassaemic major. AGC: The other baby is a carrier. The detailed pedigrees of the father and mother reveal the transmission of the thalassaemic gene, both horizontally and vertically. The genetic status of the grandparents from both sides is not available. Note the presence of the thalassaemic gene in other relatives.

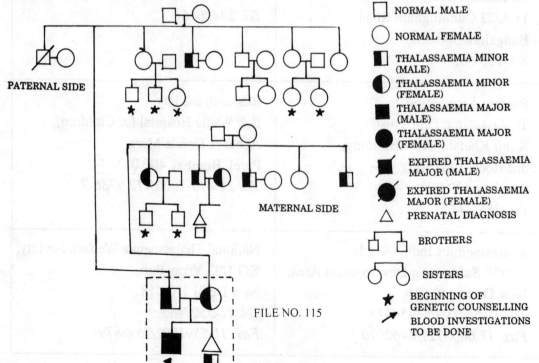

Case No. 2 (File no. 204): BGC: The two carriers gave birth to eight children. Out of these, seven are thalassaemic major, five expired soon after birth, and two are still alive. One child is a carrier and has undergone blood investigations. AGC: Prenatal diagnosis ensured the birth of a healthy child.

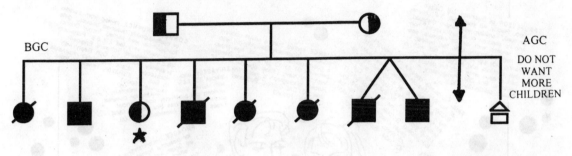

FILE NO. 204

Case No. 3 (File no. 27): BGC: Six children were born to two carriers. Of the four children with thalassaemic major, three have expired. There are two carriers who will be undergoing blood investigations. AGC: Parents do not want more children.

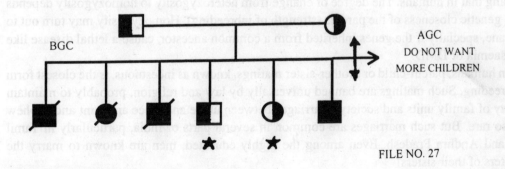

FILE NO. 27

Vinita decided to terminate her second pregnancy as the genetic tests revealed that the foetus carried the thalassaemic gene. The ensuing trauma and all that they have learnt about genetics have led the Vasvanis to contemplate their family links. Both Vinita and Vinod are distantly related and it was easy to fix their marriage. But this very approach of marrying among relatives, however distant, needs a critical review by society.

In India it is common to arrange marriages among relatives, as in the case of the Vasvanis. The worrying aspect here is that couples in these groups may possess copies of one or more analogous genes inherited from a common ancestor. This type of marriage between relatives, a type of inbreeding, is called consanguineous and the children of such marriages are said to be inbred.

In nature, inbreeding is common in self-fertilized groups of plants, like the garden peas used by Mendel. His experiments had beautifully revealed that continuous self-fertilization increased the fraction of homozygosity for certain characters; there was a corresponding decrease in the number of plants with heterozygous characters. This happens in all forms of inbreeding, including that in humans. The degree of change from heterozygosity to homozygosity depends on the genetic closeness of the parents (strength of inbreeding). Homozygosity may turn out to be a bane, specially if the genes inherited from a common ancestor, cause a lethal disease like thalassaemia or DMD.

In humans, parent-child or brother-sister matings, known as incestuous, is the closest form of inbreeding. Such matings are banned universally by law and religion, probably to maintain stability of family units and society. Marriages between uncle and niece and aunt and nephew are also rare. But such marriages are common in several parts of India, particularly in Tamil Nadu and Andhra Pradesh. Even among the highly educated, men are known to marry the daughters of their sisters!

The crux of the problem in such inbreeding is to ascertain whether the increased homozygosity of genes is for good or bad genes. One can never be sure about this, but inbreeding certainly brings out the detrimental recessive alleles (responsible for thalassaemia, alkaptonuria, phenylketonuria, albinism, cystic fibrosis, Tay Sachs, repeated abortions, low weight birth babies, etc.) which otherwise are hidden when they are present as a single copy in heterozygous carriers. The important point is to check out the detailed family history of both the partners. The presence of a deleterious allele in one of them or in any of the ancestors should alert the couple.

A family tree showing interrelationships.

Whatever be the level of consanguinity, the chances of homozygosity of genes increases among the inbred progeny. Of course, a person could be homozygous without having parents who are related.

Inherited diseases cause great pain and grief to individuals and to their families. Palliative measures, in the absence of treatments for most of these diseases, are not of much help. No wonder, stress is laid on prevention of these diseases by physicians and counsellors alike. And in this preventive strategy, screening of at-risk populations or certain groups seems to be a rational approach.

In genetic screening, routine diagnostic procedures are used for the detection of inherited diseases in a target population. Screening is generally carried out on groups of people and not on an individual basis.

SCREENING IS AIMED AT DETECTING THOSE AT RISK: CARRIERS OF A DISEASE GENE OR THOSE AFFECTED!

A good strategy is to screen the newborns. Inherited metabolic disorders, if identified at an early stage, can be appropriately treated. For instance, mental retardation associated with accumulation of phenylketonuria can be prevented if a special diet, low in amino acid, phenylalanine, is given to infants. In another instance, babies cannot synthesize the enzyme, biotinidase, involved in the metabolism of a vitamin, biotin. These babies then develop fatal abnormalities. However, they can lead healthy lives if the infants are given pharmacological doses of biotin at an early age. Screening for several metabolic disorders is often carried out by simple biochemical tests carried out on prenatal (maternal) blood, or the cord blood, or the blood/urine of the new born. Such screening, followed by remedial treatments, gives a new life to the affected individuals.

Screening for other inherited diseases, such as blood disorders, including thalassaemia, dystrophies, or Huntington's chorea, would be of great help, especially if carried out at prenatal stage—in early months of pregnancy. Biochemical and molecular tests have made screening for these diseases relatively simple. With this rationale, screening is strongly advocated for women during early pregnancy. Screening could be carried out on young couples, especially those belonging to small inbred religious groups.

Several countries, including Great Britain, Italy, Greece, and Cyprus, carry out state-supported prenatal diagnoses and termination of pregnancies, when the foetus is diagnosed with thalassaemia and other haemoglobinopathies. In fact, a majority of mothers in these countries are keen to avail of these services, more so if they have already borne one thalassaemic child. Screening of all prospective parents is a worthwhile exercise, especially those bearing children when they are above 30 years of age. If, for instance, the foetus is identified with Down's Syndrome or thalassaemia at the prenatal stage, the pregnancy could be terminated. Needless to say, large scale screening tests should be inexpensive.

In a 1997 survey of school children in Delhi, conducted to detect the presence of the thalassaemia gene, five per cent of students in one of the schools were found to be carriers of the gene. Considering this high incidence of thalassaemia, the Thalassaemic Society of India has recommended to the Ministry of Health and Ministry of Education to screen all students in schools in Delhi. According to the Society, nearly 8,000 thalassaemic babies are born annually in India. The screening blood test costs a mere Rs 120/-, a small price for avoiding life-long misery.

There are difficulties associated with large scale screening of populations. Those who oppose screening question the reliability of diagnostic tests. They also question the wisdom of branding people at a young age as carriers of the so-called 'bad genes'.

Poor understanding of genetics leads to rampant branding with resultant discrimination, especially to a person who is a carrier and not really affected by the disease. Opposition to screening in schools is often for this reason. Any programme that does so must be carried out with utmost sensitivity and confidentiality.

Economics of a screening programme in a country like India must be carefully ascertained. Funds that could help the affected should not be squandered on identifying potential carriers, often with no family histories. This view is not endorsed by all and is debatable.

'I believe that the medical benefits of screening for the most crippling inherited diseases and predispositions to certain cancers, outweigh the social dangers and moral objections; I believe that everyone has the right to be given what information he asks for about his own genetic make-up and that of his newborn children, but that no one else has a right to information about the genetic make-up of other people and their children. Genetic screening will be acceptable provided the law guarantees that exclusive right to genetic information, and the right of parents to individual choice.'

(Max F. Perutz, Nobel Laureate.)

The sheer possibilities and available prospects offered by molecular biological techniques are heady, but scary. The human genome project expected to be completed by 2003 with all the genes marked out and mapped on the 23 pairs of human chromosomes, will release a tidal wave of genetic information. While much of the information will be of great use to mankind, scientists and concerned citizens are worried about the misuse of this information.

There is little doubt, even amongst the most hard-headed sceptics, that the tools of molecular biology have enormous potential to do both good and bad. After all, we are now dealing with the DNA of man and not that of mice or monkeys. On the one hand, prenatal diagnostic tests on a few cells drawn from the foetal mass, is already determining the sex and health of newborns. The mapping of the genome and gene therapy will bring us closer to tackling several inherited diseases for which, at present, we have neither cures nor hopes.

But a host of complex questions and dilemmas are also going to crop up, including questions about invasion of our privacy, to discrimination and playing God with our genes. As even wisps of DNA get deciphered, with meanings ascribed for all the nucleotide bases, it will be rather difficult to distinguish between a normal and an abnormal gene.

Patents are being filed for big and small pieces of DNA as monetary considerations are involved. According to an official in the US Patent and Trademark Office, 'no one owns the gene in your body, investors can own the right to exploit it commercially'. Patents for DNA fragments coding for interleukin and interferon proteins involved in boosting our immune system, and erythropoietin genes needed by kidney patients, are already fetching billions of dollars for their owners! To counter this, academic scientists are pleading for putting the information about new genes first in the public domain so that patenting is somehow slowed down.

Fears have also been expressed that our genetic information will be misused by different agencies. The misuse of genetic information is not a new phenomenon. In 1965, several mental and criminal inmates were found to carry an extra Y chromosome. This resulted in hysteria and brought forth amazing demands.

'Carry out massive prenatal screening, And then abortions,
To weed out the criminals.'
'Carry out long-range studies. Track such individuals.
At their homes, schools and work places.'
'Society needs to be protected from such individuals.'

These fears were unfounded, for subsequent studies proved that nearly 96 per cent of people with a XYY pattern lead normal lives. What is the guarantee that such hysteria will not recur?

Ownership and availability of one's genetic portfolio is going to raise a set of questions to which there are no answers yet. Who should possess an individual's genetic data? Should these data be stored in hospitals and laboratories? Should the information be available to insurance companies and to employers? Or, should it be confidential? Should the decision to divulge one's own genetic data to different agencies or to the family members, be left to an individual? Perhaps, the individual may not even want a peep into his genetic legacy.

If the data become freely available, managements may wish to employ only those who are perfectly healthy and refuse jobs to the carriers of disease-gene, on the slightest pretext. One can well imagine the rising tide of the unemployed since most of us carry one rogue gene or the other. Indeed, today those testing HIV positive are finding it difficult to retain their old jobs or getting new ones. In addition, they have to pay higher premiums to medical and life insurance companies. Discrimination of this sort can be stopped only if prohibited by law, but the potential for intolerance and prejudice cannot be ruled out.

It seems we are just a tight bundle of molecules.

Living organisms are NOT just an aggregate of molecules. The highly sophisticated organization of chemical molecules leading to their complex structures and functions—at individual and group levels—marks them out from a mixture or group of chemicals. At the most simplistic level, molecular organization in living organisms could be compared to a beautiful musical composition, say by Pandit Ravi Shankar. The *raga* is composed of several sets of musical notes(= molecules) put together to produce beautiful music.

At a personal level, the genetic information can be of great help to an individual. For instance, if identified with a cancer gene, a healthy lifestyle may ward off the disease by a few years or even completely. The information may help to put one's life in order. But then the entire experience may turn out to be too traumatic and painful for an individual. 'Who wants to know when I will die', is a common comment made to genetic counsellors.

The rising tide of DNA wave will label all of us either as sick or healthy. Thus, 'branded' individuals could develop an inferiority complex about their birth. Saddled with labels, followed by possible elimination of 'rogue' genes, there is the potential of tipping the scale towards the creation of a race with superior genes. The eugenic movement in the 1900s, leading to extermination of millions in Germany, was guided by this very premise!

In the future, one cannot rule out the possibility of gene therapy, being tried out on germ cells. As James Watson, the most vocal champion of the HGP said, 'The objective should not be to get genetic information per se, but to improve life through genetic information.' The plethora of questions associated with HGP will multiply as the genetic apparatus gets increasingly untangled. The proponents of the HGP had anticipated the associated problems and hence nearly 3 per cent (and now 5 per cent) of the HGP budget is devoted to the task of addressing 'ethical, legal and social implications' (ELSI) of the public use of genetic information, in addition to finding ways for information to be beneficial to society. In fact, scientists are actively guiding and propelling the scientific findings and developments, keeping in mind the ethical and other related issues.

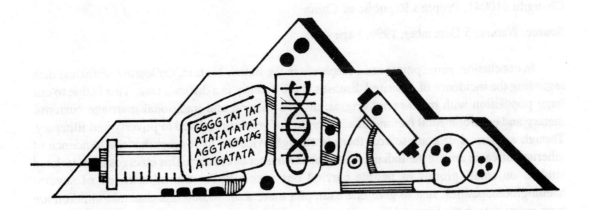

Ethical issues associated with human genetics elicit varied responses. All societies desire to reduce the incidence of inherited diseases in their population. The approaches, however, vary from being persuasive (as in the West and in India) to aggressive.

CHINESE ETHICS

Since the Chinese Maternal and Infant Health Care Law, previously called the 'eugenics and health protection law', first proposed in December 1993, and finally promulgated on 27 October 1994, it has provoked widespread concern in the international scientific community. What is the attitude of Chinese geneticists towards ethical issues, including the controversial 'eugenics law'? A national survey, funded in part by the Ethical, Legal and Social Implications Branch of the US National Centre for Human Genome Research, was conducted among 402 geneticists from 30 provinces and autonomous regions in China.

The survey showed that 95 per cent of Chinese geneticists agreed that 'people at high risk for serious disorders should not have children unless they use prenatal diagnosis and selective abortion'; 90 per cent agreed with the statement that 'an important goal of genetic counselling is to reduce the number of deleterious genes in the population'; and 90 per cent called for ethical guidelines for genetics practice and research in China. Eighty-nine per cent supported current Chinese laws on abortion for genetic abnormalities and non-medical indications.

More than half opposed sex selection by any means; 55 per cent thought that gene donors should have a right to share in the profits from commercialized tests or treatments developed using their genes; and 50 per cent thought that public education on genetics should be the top priority of the Chinese government health budget. This is the first survey to provide a basis for international discussion of ethics and genetics in China.

Xin Mao, Division of Genetics
West China University of Medical Sciences,
Chengdu 610041, People's Republic of China

Source: *Nature,* 5 December, 1996, page 404.

In conclusion, some points need emphasis in the Indian context. Collecting statistical data regarding the incidence of inherited diseases in the country is a difficult task. This is due to our large population with multi-racial origins, diverse cultural and traditional marriage customs, dietary and environmental factors, in addition to the problems related to poverty and illiteracy. Though numerical data are scarce, there is enough evidence to suggest that the incidence of inherited diseases among the Indian population is a matter of concern. This emerges from isolated studies on small groups of people carried out specially for the inheritance of several haemoglobinopathies. Just to give one example, more than 10,000 thalassaemic children are born every year in the country! It is apparent that strategies to prevent inherited diseases should be high on the agenda of public health policy.

Education and awareness about human genetics and diseases can play an important role in preventing inherited diseases. Even the highly qualified are ignorant of the science of genetics. In fact, everyone needs to understand basic genetics. For instance, there are differences between a 'carrier' of an abnormal gene and a person being affected with a genetic disease; between genetic markers for a disease and the disease-gene itself. It is imperative to discern these intricacies. Equally important, individuals should also learn to differentiate between genetic probabilities and genetic certainties. In the absence of this understanding, people are likely to adopt attitudes of naive genetic determinism and think only in terms of good and bad genes, or that genes control behaviour.

Proper understanding of genetics can help people take right decisions when confronted with complex situations likely to arise as the Human Genome Project concludes. People need to be informed when difficult personal decisions, such as 'adoptions versus abortion', have to be taken. The situation demands education of the general public and of those who are supposed to impart health care to the public. There is a strong case for introducing human genetics at the school and college levels, for students of all subjects. This book is a modest step in this direction.

Appendix

List of some genetic clinics and counselling centres in India

1. Genetic Clinic, All India Institute of Medical Sciences, Ansari Nagar, New Delhi 110029
2. Department of Pathology, Vivekananda Institute of Medical Science, Ramakrishan Mission Seva Pratishthan, Calcutta 700026
3. Human Genetics Unit, University of Calcutta, Ballygunj Circular Road, Calcutta 700019
4. Institute of Immunohaematology, Parel, Mumbai 400012
5. Research Society, B. J. Wadia Hospital for Children and Institute for Child Health, Mumbai 400012
6. Department of Paediatrics and Biochemistry, B. J. Medical College, Pune 411001
7. Genetics Unit, Bhagwan Mahavir Medical Research Centre, Hyderabad 300004
8. Human Genetics Department, University of Delhi, Delhi 110007
9. School of Sciences, Gujarat University, Ahmedabad 380009
10. Institute of Medical Sciences, Banaras Hindu University, Varanasi 221005
11. Institute of Genetics and Hospital for Genetic Diseases, Osmania University, Hyderabad 500016
12. Centre for Genetic Disorders, Guru Nanak Dev University, Amritsar 143005
13. National Cancer Registry (ICMR), Tata Memorial Hospital, Mumbai 400012
14. Department of Human Genetics and Reproductive Biology, Bombay Hospital, Mumbai 400020
15. Department of Paediatrics, KEM Hospital, Mumbai 400012
16. School of Biological Sciences, Madurai Kamaraj University, Madurai 625021
17. Department of Paediatrics, Medical College, Trivandrum
18. R. P. M. College, Uttarpara, West Bengal
19. Department of Anthropology, Andhra University, Waltair 530003
20. Smt Motibai Thakersey Institute for Research in the Field of Mental Retardation, Sewri Hill, Sewri Road, Mumbai 400033
21. Department of Anthropology, University of Delhi, Delhi 110007
22. Department of Genetics, Osmania University, Hyderabad 500007
23. Division of Biochemistry, Department of Chemistry, University of Poona, Pune 411007
24. Cell Biology Division, Gujarat Cancer and Research Institute, Ahmedabad 380016
25. Cancer Research Institute, Mumbai 400012
26. Genetic Division, Grant Medical College and J. J. Group of Hospitals, Mumbai 400008
27. Genetic Division, Sion Hospital, Mumbai 400022
28. Department of Genetics, Sir Ganga Ram Hospital, Rajinder Nagar, New Delhi 110060

References

1. *Peoples of India: Some Genetical Aspects;* XV International Congress of Genetics, New Delhi, ICMR, New Delhi, 1983.
2. *Human Genetics: A Modern Synthesis* by Gordon Edlin; Jones and Bartlett Publishers, Boston, 1990.
3. *Genome* by Jerry E. Bishop and Michael Waldholz; Touchstone, New York, USA, 1990.
4. *Basic Human Genetics* by Elaine Johansen Mange and Arthur P. Mange; Sinauer Associates, INC. Publishers, Sunderland, Massachusetts, 1994.
5. *The Human Genome Project: Deciphering the Blueprint of Heredity;* Edited by Necia Grant Cooper; University Science Books, CA, USA, 1994.
6. *Molecular Biology of the Cell* by Bruce Bray *et al.*; Garland Publishing, Inc., NY, 1994.
7. Highlights of Tribal Health Research Under Indian Council of Medical Research: ICMR Bulletin, March-April, 1996.

616
Mah

Mahajan, Bakhtaver
S.

New biology and
genetic diseases.